AF471323

LIGNUM

A History

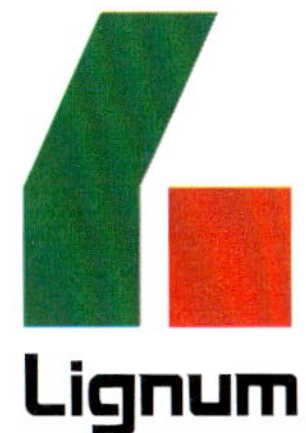

Lignum

LIGNUM

A History

KEN DRUSHKA

Published by:
Lignum Ltd.
Suite 1200, 1090 West Georgia Street
Vancouver, B.C. Canada V6E 3V7

Cover design, page design and composition by Roger Handling, Terra Firma Digital Arts. Editing by Daniel Francis.

Printed in Canada.

Canadian Cataloguing in Publication Data

Drushka, Ken
           Lignum

ISBN 0-9731084-0-1

  1. Lignum (Firm) 2. Sawmills--British Columbia--History. I. Lignum
(Firm) II. Title.
HD9764.C44L53 2002   338.4'76742'09711   C2002-910665-6

Photograph credits:
p. 11, Golden District Historical Society; p. 26, Chase & District Museum and Archives; p. 28, 69, Author's collection; 45, David Ainsworth; 57, 65, 66, 67, 131, San Jose Logging; 110, 111, Peter Cardew; 133, Jackpine Group; all others, Lignum Ltd.

Preceeding page: Aerial view of Williams Lake planer mill, 1950.

Right: A new forklift was part of the technological change that transformed the face of the industry in the 1950s.

CONTENTS

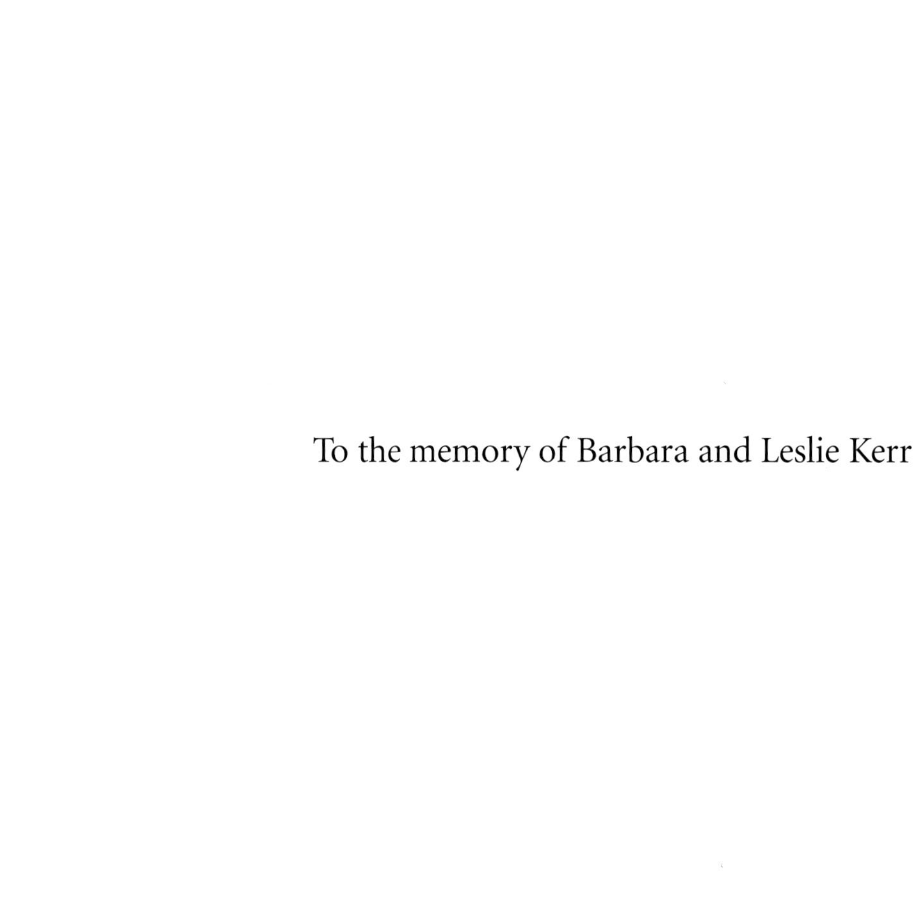

To the memory of Barbara and Leslie Kerr

FOREWORD

L ignum was created by my father, Leslie Kerr, in 1946 with a small planing mill in Quesnel and a sales office in Vancouver. During the 26 years he led the company, its production division was relocated to Williams Lake and expanded to include a sawmill. At various times, other mills were purchased or built, but eventually the decision was made to focus attention on one mill at Williams Lake and make it the most productive manufacturing facility of its kind.

Throughout the early period, much of the corporate focus was on the acquisition of a secure timber supply in the Cariboo-Chilcotin region. Since that time, the company has assumed the responsibility for management of the forests which provide that supply.

A key element of my father's business ethic was the establishment and maintenance of close working relationships with individuals and other businesses. The company's early success was built upon the relationships between the operators of bush mills who provided the rough lumber that was finished in Lignum's planers and sold in Canada and abroad by the Vancouver sales office.

Over the past 30 years, additional relationships have been forged in the Cariboo with logging contractors, suppliers and remanufacturers, and in the marketplace with a range of customers.

From the outset, the success of Lignum has depended on the people employed by the company. This includes those who established the first mills and remained to rebuild them in response to new needs and opportunities, creating with their efforts one of the most efficient and productive facilities in the country. It also includes a sales staff which has often led the way in the distribution and marketing of forest products. And, it includes a strong management team which has guided the company through some of the most challenging times the industry has faced.

For some time now, it has occurred to me that the unique story of Lignum and the people who built it should be told. I wanted this to be a candid story, one that acknowledged some of our failures, as well as our accomplishments and our triumphs. I wanted to leave an account of Lignum's creation for the next generation of loggers, mill workers, lumber traders and managers who will remake what we have built into a company that suits the circumstances of their time.

The principles upon which Lignum was founded, and upon which it has grown and thrived have created a unique company that has achieved many successes during its 56 years of operation.

I am very proud of the accomplishments of the people who built Lignum. We could never have done it without each other.

Jake Kerr
Chairman and CEO, Lignum Ltd.

Cariboo-
Chilcotin
Region

New Hazelton
Granisle
Smithers
Telkwa
Houston
Burns Lake
Francois Lake
Fraser Lake
Ootsa Lake
Vanderhoof
Prince George
Red Rock
Stoner
Hixon
Strathnaver
Wells
McBride
Mt Robson
Tête Jaune Cache
Valemount
Nazko
Quesnel
Barkerville
Kersley
Likely
Blue River
Marguerite
McLeese L
Horsefly
Soda Creek
Anahim Lake
Nimpo Lake
Williams Lake
150 Mile House
Eagle Creek
Clearwater
Bella Coola
Chilanko
Forks
Alexis Creek
Springhouse
Lac la Hache
Canim Lake
Riske Creek
Alkali Lake
100 Mile House
Little Fort
Kleena Kleene
Tatlayoko Lake
70 Mile House
Nemaiah Valley
Barrière
Mt Waddington
Clinton
Heffley Creek
Chase
Tappen
Sicamous
Cache Creek
Salmon Arm
Lillooet
Logan Lake
Kamloops
Pemberton
Lytton
Merritt
Whistler
Kelowna
Garibaldi
Squamish
Vancouver
Hope
WASHINGTON
VANCOUVER
ISLAND
Victoria

Bulkley R
Takla L
Tremb
Babine L
Morice L
Stuart R
Oostsa L
Nechako R
Willow R
Bowron R
McGregor R
Tahtsa L
Chilako R
Fraser R
Whitesail L
Canal
Eutsuk L
Tetachuck L
West Road R
Fraser R
Quesnel R
Quesnel L
Hobson L
Kinbasket L
Dean R
Horsefly L
Kittlope R
Murtle L
North Thompson R
Bella Coola R
Chilcotin R
Mahood L
Dean Channel
Charlotte L
Chilko R
Canim L
Columbia R
Burke Channel
Taseko L
Qwikeno L
Klinaklini R
Tatlayoko L
Adams L
Sound
Caution
Homathko R
Chilko L
Bonaparte R
Shuswap L
Uppe
Knight Inlet
Fraser R
Kamloops L
Bridge R
Sound
Bute Inlet
Anderson L
Thompson R
Nimpkish L
Lillooet R
Nicola R
Okanagan L
Toba Inlet
Cook
Powell L
Jervis Inlet
Lillooet L
Fraser R
Low
Arrow
Kyuquot Sound
Harrison L
Nootka Sound
Pitt L
Similkameen R
Great Central L
Strait of Georgia
Barkley Sound
Cowichan L
Cape Flattery
Juan de Fuca Strait

INTRODUCTION

Lignum's award-winning office at the Williams Lake mill.

This is a remarkable story about the building of one of British Columbia's most successful forest companies. In a little more than 50 years Lignum Limited grew from a small planing mill in Quesnel to become one of the most efficient and profitable lumber producers in Canada.

What is even more remarkable is that this private, family-owned firm thrived and grew in an era of corporate takeovers, consolidation and growth-by-acquisition—not by adopting these tactics, but by marching to the beat of its own drums.

At the heart of Lignum lies an idea, or a concept, articulated and followed by the company's founder, Leslie Kerr. This was not so much a plan as an attitude. It was a way of doing business that stressed the importance of establishing and maintaining personal relationships, both within the company and in the company's dealings with its business partners. It is an idea and an attitude that has guided his descendants and those who have built upon the foundations he laid.

Leslie Kerr was a visionary. Fifty years ago he saw a way for the Cariboo forest industry to evolve that would avoid conflict and the displacement of those already established in the business.

He was not the sort to go about loudly proclaiming his ideas. The only clear record he ever left that describes them was in response to questions at a Royal Commission on the future of the province's forest industry.

Leslie saw beyond the immediate economic opportunities offered by the lumber industry of the late 1940s. He recognized the value of the province's timber and, even beyond that, the enduring economic worth of the forests that provide that timber.

In recent years, Lignum has prospered by accepting a timber supply spurned by others and manufacturing it into a premium product for the North American market. Its business strategy has enabled it to survive, relatively unscathed, through a particularly turbulent period of industrial unrest. Unlike many of its less-fortunate competitors it has done this without the need to lay off its workers, temporarily close its mill, or assume a crippling debt load.

Today, Lignum is among a handful of BC forest companies that looks to the future as a time of opportunity and promise. Its forest-management plans anticipate not only the company's needs but the requirements of non-timber users and values as well. Its production facilities are as finely tuned as any in the business. And its marketing strategy foresees rather than follows changing demands.

No one at Lignum believes for a moment that it stands as a model for success and emulation. There is a modesty about the company and the 450 people who operate it. They strive to do what they feel is right for themselves, in the circumstances they encounter. They do not presume to advise others.

For an outsider, though, it is hard to avoid the thought that the ideas made concrete by Leslie Kerr, and refined by his sons and their co-workers, have application well beyond the precincts of company operations. In an age of economic uncertainty and corporate upheaval, Lignum's status and the story of how it got to where it is today is worth careful consideration.

CHAPTER 1

MIGRATION TO THE CARIBOO

T HE CARIBOO-CHILCOTIN REGION OF CENTRAL BRITISH COLUMBIA STRADDLES THE FRASER RIVER BETWEEN THE COAST MOUNTAINS ON THE WEST AND THE CARIBOO MOUNTAINS TO THE EAST. THE REGION BEGINS AT AN ILL-DEFINED SPOT SOMEWHERE NORTH OF KAMLOOPS AND EXTENDS TO AN EQUALLY ILL-DEFINED LOCATION SOUTH OF PRINCE GEORGE.

Until well after the end of World War II, the economy of the region remained in a pre-industrial state, made up for the most part of ranchers and cattlemen, and residents of the small communities that served them. Meanwhile, the rest of the province had been acquiring the appurtenances and infrastructure of a modern industrial society.

Construction of trans-continental railway lines to the north and south of the region in previous decades had precipitated industrial development in many other parts of the province. Completion of the Canadian Pacific Railway in 1885, and the subsequent extension of branch lines throughout southeastern BC, stimulated the building of industrial enterprises along its right of way.

The Columbia River Lumber Co. sawmill at Golden, which operated from 1900 to 1927, was typical of large mills built along main transportation corridors decades before stationary mills were established in the Cariboo-Chilcotin region.

Large sawmills built to service the prairie market were established from Kamloops east to Golden, and throughout the east and west Kootenays, where mining developments also appeared. An agricultural economy flourished in the Okanagan, and Kamloops became a major shipping centre for lumber, mineral concentrates and cattle.

Completion of the Grand Trunk Pacific line to Prince Rupert in 1914 opened the timber-rich upper Fraser River valley east of Prince George, where a score or more large sawmills were built. West of Prince George, all the way to saltwater, a flourishing tie and pole industry developed which financed much of the agricultural development of the Bulkley Valley.

On the Coast, especially along the eastern shore of Vancouver Island and the Lower Mainland around Vancouver, the period after World War I was one of rapid industrial development. Driven by the extraction and utilization of natural resources, the booming BC economy was of sufficient size to spawn its own manufacturing offshoots, further increasing growth. By the end of the 1920s, more than half of BC's half-million residents lived in and around Vancouver, a city well deserving of its designation as the Big Smoke by residents of the provincial hinterlands.

Over the course of a two- to three- year period, beginning in late 1929, this energetic young economy collapsed amid the general global depression that undermined the export markets upon which it depended. Social and political upheaval followed, with riots and pitched battles between unemployed industrial workers and the forces of law and order waged in the streets of Vancouver. To the widely scattered residents of the Cariboo-Chilcotin these were distant events, not as far

removed as those in eastern Canada or Europe, perhaps, but of little immediate consequence.

There were, then as now, two populations in the region. Several thousand aboriginal people, whose ancestors arrived following the end of the last ice age about 10,000 years ago, lived on reserves scattered throughout the region. The lands east of the Fraser were occupied by the Shuswap (Secwepemc) people, while the Chilcotin (Tsilhqoqt'in) lived west of the river. The Shuswap, who had a long experience with the newcomers, co-existed with them peacefully. The more remote Chilcotin maintained hostile relations with the early fur traders and miners until the conclusion of the so-called Chilcotin War in 1864 when 143 workers building a trail from the Coast to the Cariboo gold fields were killed and five natives were hanged in retaliation. The populations of these two peoples were devastated by a smallpox epidemic in the 1860s, and by the 1930s descendants of the survivors had settled in a number of permanent villages on reserves located throughout the region.

The Cariboo-Chilcotin's non-aboriginal population, by the outbreak of World War II, was in its third phase of development. The first wave of outsiders to settle in the area were traders employed by the Montreal-based North West Company who built Fort Alexandria on the east side of the Fraser, about 320 kilometres north of present-day Soda Creek, in 1814. It was relocated a couple of times, eventually at a site on the west side of the river.

The next influx of newcomers began in 1859 when gold was discovered in the Cariboo Mountains. Thousands of gold seekers poured into the area, most of them travelling up the Fraser to its junction with the Quesnel River, then east to the centre of mining activity at Barkerville. The colonial government constructed a wagon road from the head of navigation on the lower Fraser at Yale to the gold fields in 1862-64.

The frenzied search for gold spun off a variety of other developments in the region undertaken by people whose primary interest was not gold. Principal among these were cattlemen who

Miners at the Neversweat Mine at Williams Creek during the Cariboo gold rush. (BC Museum of Mining)

drove herds of beef from Oregon to supply the gold miners. Many of them obtained grazing leases in the extensive grasslands on both sides of the Fraser and began building herds. When the gold rush ended, many of the ranchers remained, driving their cattle for sale and shipment at Ashcroft after the CPR reached that community on the southern edge of the Cariboo in 1884.

These cattlemen were of mixed origins. Some of them, such as William Pinchbeck, who located his ranch on the present-day Williams Lake Stampede grounds, were well-educated, well-financed English immigrants. Others, like Henry Felker, who established a ranch near Felker Lake, and Herman Otto Bowe, who founded the Alkali Lake Ranch (the first beef cattle ranch in BC), came from Germany, via the US. Joseph Dussault, from Quebec, hauled freight with oxen on the Cariboo Wagon Road before settling west of Williams Lake. Augustine Boitaino, from Italy, also packed freight before building the first ranch in the Springhouse district.

These were tough, independent pioneers who thrived living in semi-isolation. They shaped the character of the civil society that began to evolve in the region. The settlements that grew to serve them and the resident native population reflected the characters of both peoples.

The cattle industry dominated the region's economy until well into the 20th century. By the end of World War II it was still the region's primary economic activity, and was centred in Williams Lake, a small town located a few kilometres east of the Fraser River.

Sitting more or less at the geographical centre of the Cariboo-Chilcotin, Williams Lake epitomized the region. It began with the establishment of a gold commissioner's office in 1860 near the large Indian settlement of Columneetza on the shore of a small lake at the mouth of the San Jose River. A courthouse and jail were built, but then the community died when it was bypassed by the Cariboo Wagon Road. For several decades the only activities in the area were associated with William Pinchbeck's ranch and the Sugar Cane Reserve, at the head of the lake.

The settlement came to life again in 1919 when the Pacific Great Eastern railway connected the region to the Coast at Squamish. The railway company bought the Pinchbeck ranch and laid out a town site, with Railway Avenue parallelling the rail line, and Oliver Street—named after BC premier John Oliver—striking off up the hill from the depot. A dusty cow town began to grow out from these two thoroughfares.

Thirty years later, Williams Lake was still no thriving metropolis. The town had grown a few blocks up Oliver Street, and a few blocks each way along Railway, renamed Mackenzie Avenue after the first merchant to build a store in the community. The Cariboo Highway, an unpaved road, virtually impassable in wet weather, came through the town along Mackenzie Avenue on its way north. It passed a small cemetery on the northern outskirts of town.

For most of the week, Williams Lake was a quiet, sleepy place. Saturdays were more lively as the hotels filled with ranchers and ranch hands in for supplies and a night of fun. Twice a year, the town erupted in activity.

An annual cattle sale took place every October, with ranchers driving their herds in for sale, bringing along their families and ranch hands. By the 1940s Williams Lake had become the largest cattle shipping centre in the province. Auctions went on for a week, beginning early in the morning and ending in time for the nightly celebrations, which included dances, banquets, casino nights and an unending round of private parties.

The banks staged huge parties, beginning with discreet, invitation-only affairs at the managers' houses, that eventually became so big they were staged in the banks themselves.

Above: Until well into the 20th century the cattle industry dominated the Cariboo-Chilcotin economy.
(Vernon Museum)

Right: Mackenzie Avenue in Williams Lake in the 1950s, facing south from near the Lignum mill yard. The old Imperial Oil warehouse is on the right. At this time, the street was part of the Cariboo Highway.

The other annual event was the stampede. The first Williams Lake Stampede was staged in 1919, the year the railway arrived. It was a hastily-organized event held at the old Pinchbeck ranch on the site of the present stampede grounds. Word spread throughout the region that a big event of some sort was being organized and people appeared from all directions—ranchers, natives, settlers—everyone who lived within a couple of days travel of the townsite. They camped in the fields and spent the days watching riding, roping and racing contests, dancing the nights away in the ranch house.

It was so much fun, a more organized version was planned the following year and, except for a few years during World War II, it has taken place ever since. The stampede was a roaring success from the outset and within a decade was known throughout North America and abroad as the most authentic event of its kind.

It was also known as a wild and woolly affair that reflected the hard-working, hard-drinking and fun-loving residents of the region. One event that contributed to this reputation was the

February 11, 1920. The store building being erected by Messrs. Fraser and Mackenzie is about completed and the citizens of Williams Lake organized a dance. People began to arrive Thursday and by Friday gambling was in full swing. The big show began at midnight when a train came in bringing a number of outsiders and a generous supply of liquor. It was not long before nearly everybody was pretty well under. One man had to be ejected from the building and a crowd followed him out when we locked the door and let them fight it out amongst themselves. Nobody was killed.

Louis P. Dallaire, first manager of Williams Lake branch of the Bank of Commerce, in a report to his head office.
In Cariboo-Chilcotin: Pioneer People and Places by Irene Stangoe.

Mountain Race. First staged in 1922, it was the final stampede competition for 30 years. The race began on Fox Mountain, above the present Overlander Hotel. The mile-long course began with a jump off a ledge, dropped precipitously down a steep gully through fallen trees and loose rocks, and out onto the flats, ending on the stampede race track. Huge pile-ups were normal, as were injuries to horses and riders.

A second event that typified the stampede for some and, for others, gave it a bad name, was the legendary Squaw Hall. It began as a dance hall for natives in 1947, but quickly became one of the most popular events of the stampede—with 7,000 to 8,000 people attending—until it was terminated in 1975.

In its heyday Squaw Hall consisted of a wooden dance floor, surrounded by four walls, and open to the sky. In its later years, part of the fun involved the hurling of empty beer bottles over the walls onto the dance floor—and anyone who happened to be in the way. Dancers shuffled their way through a layer of broken glass, while the orchestra played on, ensconced inside a protective chicken-wire cage.

With the second great war over and the provincial economy beginning to boom, life in Williams Lake continued much as it had for the previous 25 or 30 years. At most, a thousand residents lived there. It still served as the centre of the Cariboo-Chilcotin ranching community, which had grown slightly over the years and still retained its rough-edged character. Stray cows could be seen wandering along Oliver Street, and bar-room brawls were common occurrences.

All this was about to change. The agent of this change—given the character of the location he had selected for his new enterprise—was as unlikely a person as could be imagined. One day in early April, 1948, an elegantly dressed man in a well-tailored sports jacket, immaculately pressed pants, and highly polished shoes disembarked from a PGE passenger train, carefully picked his way through the mud of Mackenzie Avenue, and walked up the wooden sidewalk along Oliver Street.

He had come to announce, in a refined European accent, the launching of a new industrial enterprise for Williams Lake and the Cariboo-Chilcotin. His origins were as exotic as many of the region's first settlers, who had arrived almost a century earlier.

Not all parts of the world came through the Depression and war as unscathed as the Cariboo-Chilcotin. For most of the 1930s and until the end of the war Europe was in chaos. As fascism triumphed in country after country, a vast movement of people began, some fleeing for their lives, others searching for a place to live a life free of totalitarian governments.

Leslie Kerr with his two sons, Tim and John, on the streets of Vancouver in the early 1950s.

The rough and ready Williams Lake Stampede was considered the most authentic
event of its kind during the mid-1900s.

Top: Bill Twan, chariot racing at the Stampede in the early 1950s.
(Twan Family Archives)

Bottom: Downtown Williams Lake in the same era.
(Twan Family Archives)

*Top: A Czechoslovakian sawmill whose lumber Leslie sold in Europe in the
1930s, with lumber stacked for drying in foreground.*

Bottom: Leslie Kerr in Europe in 1937.

Among the most well equipped for this great exodus were those who worked in the European timber trades. This business, consisting primarily of the production and marketing of lumber, was centuries old and comprised one of the largest economic sectors of the European economy. By the 1920s it had evolved into an integrated network of timber growers, manufacturers of lumber and other forest products, wholesalers, retailers and final consumers. It reached throughout Europe, with tentacles around the world. Typically, firms engaged in the lumber trade were family-owned operations that had developed over several generations.

One of the most successful of these firms, founded in Czechoslovakia during the late 1800s, was owned by the Koerner family. At the end of World War I, under the direction of a new generation of Koerner brothers—Leon, Otto and Walter—the family, which operated lumber mills, entered the wholesale end of the business, setting up branches in several European countries, including Great Britain. They prospered, and Leon was instrumental in the formation of the European Timber Exporting Congress, a cartel established to maintain sales quotas for member nations in order to keep prices high. He remained in Prague to oversee the family firm, while Otto established a branch in Paris and Walter did the same in London. This placed the family at the centre of the European and, consequently, the global timber trade.

Another family-owned firm, which manufactured plywood, had been founded by the Bene family in Budapest in 1834. By the early 1930s, John Bene, who was born in Vienna in 1910, was being groomed to take over the business. After graduating from university with degrees in mechanical and electrical engineering in 1932, he worked in different parts of the trade in several European locations and formed an association with the Koerner brothers. In 1936 he visited Vancouver.

There were many similar operations of greater or lesser prominence. After World War I, the Kardos family operated a sawmill in the part of Yugoslavia that formerly made up part of Hungary. Much of their production was sold by a Hungarian family from the same area, the Kerrs, who had been in the wholesale lumber business for several generations. In 1931, 21-year-old Laci Kardos had been sent to France, where he purchased a sawmill to enable the family to escape the war already building in eastern Europe. Shortly after his arrival, he was joined in Paris by Leslie Kerr.

Leslie was born in Hungary in 1906, and educated in France and Vienna, where he obtained a degree in economics and did a brief stint teaching. From there, he went to work in Paris selling lumber, partially for the Koerners, and also for his family firm selling the Kardos family's lumber. One of his chief pleasures during this period was gambling at Monte Carlo. While in Paris, he and Bene became close friends.

Their involvement in the international timber trade gave the members of these Eastern European families a clear understanding of the dangers involved in the political events that unfolded after World War I. It also provided them with the mobility to escape the worst of these events.

In 1938, as the Nazis began their march through Czechoslovakia, Leon Koerner decided it was an opportune time to embark on a six-month world tour with his wife Thea, a former actress in the Prague theatre. On Bene's recommendation, they visited Vancouver in February, 1939, where Thea came down with mumps. By the time she recovered, war had been declared and Europe began its collapse into chaos.

The Koerner brothers scrambled to remove what assets they could from Europe. With their capital, Leon purchased a defunct sawmill in New Westminster and reopened it under the name Alaska Pine Co. It specialized in cutting western hemlock, which Walter sold in Britain under the name Alaska pine.

Otto, along with many of the family's associates and other members of the European lumber trade, joined Leon in Vancouver. Some of them went to work at Alaska Pine, while others such as John Bene, who had been able to escape the Nazi onslaught with enough of their liquid assets, went into business on their own.

Bene arrived in Vancouver in 1938 and founded Pacific Veneer Co. in New Westminster, along with two other refugees, Poldi Bentley and John Prentice, brothers-in-law who had been engaged in the textile business. They escaped with a small portion of their capital and, pooling it with Bene, seized an opportunity to fill the British need for aircraft and marine grades of plywood. From this base they expanded quickly into lumber production and logging, all of which they eventually reorganized under the name Canadian Forest Products in 1947.

Leslie came to Vancouver as part of this exodus and went to work as plant superintendent at Alaska Pine when it opened. He lived in New Westminster with a group of young men like himself who shared a love of poker and an appreciation of classical music. This latter interest took him to a symphony concert in Vancouver one evening featuring Barbara Custance as guest pianist.

She was a local musical prodigy, born in Burnaby and sent off to London as a child to study piano. In the 1930s, she played throughout Europe and, after returning to North America, settled in New York to pursue her career. That night she was in Vancouver to play before an appreciative hometown audience. After the concert, Leslie managed to obtain an introduction and in 1943, to the dismay of Barbara's mother, they were married.

The young couple easily fit into the new expatriate European colony that was already having an impact on the cultural life of Vancouver, and soon, on other parts of BC. Most of these people were highly educated, many with post-graduate degrees. Included in their ranks were internationally known intellectuals, painters, actors and musicians. They added a new element to the sparse Vancouver cultural scene.

As they became established in the forest industry, these newcomers also added enormously to the strength of the province's business community. Their familiarity with, and connections to, the European lumber trade opened up new markets for BC products. And the knowledge they brought with them, such as the Bene family's long experience with plywood manufacturing, was a valuable addition to the province's industrial capability.

Many of the new arrivals had strong entrepreneurial inclinations. Once established in Vancouver they struck out on their own. John Bene, for instance, left Pacific Veneer in 1944 and formed Western Plywood, with a plant in Vancouver and timber rights to a supply of birch logs in the Cariboo. Within five years, Western Plywood had four plywood plants, including one in Quesnel, two sawmills, a planer mill and a distribution network.

In the summer of 1945, with the war drawing to an end, Leslie Kerr followed Bene's example, leaving the security of a senior position at Alaska Pine to set up his own business. With financial backing from Ronald Graham, a Vancouver lumberman, he bought a small sawmill called Garibaldi Sawmills on the PGE line near Whistler. Lumber from the mill was shipped on the PGE to Squamish and then barged to Vancouver. The amount of reloading involved in this operation made the business financially marginal. The first winter, 2.5 metres of snow accumulated in the area, shutting down the mill for months.

Top: Garibaldi Sawmills near Whistler, 1945. This was Leslie's first business venture in British Columbia.

Bottom: The crew at Garibaldi Sawmills, 1945.

Leslie's first venture into the BC lumber business was a bust. For the next few years, between bearing children, Barbara supported the family with her musical abilities, giving lessons and concerts. At one point, fearing for her daughter's welfare, Barbara's mother obtained a job for Leslie driving a bus at BC Electric.

Realizing desperate measures were required, Leslie plunged back into the lumber business. In February, 1946, he incorporated a new company, Lignum Ltd., issuing himself the full complement of 4,000 shares, and began wholesaling lumber from an office on Pender Street. Then, with a mortgage on the family house in Kerrisdale, and with additional financial backing from Graham, he headed for the Interior, based on an understanding with his friend John Bene.

At Quesnel Leslie bought a one-third interest in a small sawmill company, Cariboo Birch Mills, established a year earlier by Fred Drezet and Gordon Pearson, that cut birch logs that were unsuitable for veneer. Pearson was a veteran Cariboo bush mill operator and Drezet was a skilled sawyer from Switzerland. When Leslie bought into the operation in May, 1947, they changed the name to Quesnel Sawmills.

It turned out to be the kind of operation the dapper, one-time gambler from Monte Carlo was looking for. And, although none knew it at the time, it was the beginning of what one day would become one of the province's most successful forest companies.

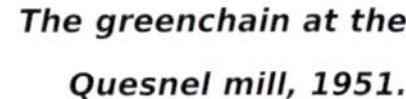

The greenchain at the Quesnel mill, 1951.

Leslie Kerr in the Cariboo, late 1940s.

CHAPTER 2

LAYING THE FOUNDATIONS

WHEN LESLIE KERR VENTURED INTO THE CARIBOO LUMBER INDUSTRY, THE REGION, ALONG WITH THE REST OF THE WORLD, WAS ON THE VERGE OF ENORMOUS CHANGE. IT WAS THE BEGINNING OF A POST-WAR ECONOMIC BOOM THAT WOULD LAST, WITH A FEW MINOR UPS AND DOWNS, FOR MORE THAN 30 YEARS.

For Canadian resource industries, this boom created many market opportunities in the United States, which had reached the limits of its domestic supplies in several sectors, including forest products. It was the beginning of a major shift in Canada's lumber markets, from Britain and Europe to the US.

Eastern Canadian sawlog supplies had been depleted by the end of World War I and the forest industries there refocused on pulp and paper production. American demand for softwood lumber, as well as pulp and paper, shifted to British Columbia.

After the war, a great movement of people occurred. It was a time of mobility between countries, within countries and within the province. A decade and a half of depression and war had created a pent-up desire on the part of many people everywhere to seek their fortunes elsewhere. The Kerrs, Koerners, Bentleys and Benes were a part of this great migration that ended up in BC, as were others from Europe, the US and the rest of Canada. People saw opportunities in the relatively undeveloped, resource-rich province and arrived in pursuit of them.

The post-war era was one of prodigious technical innovation in North America. The massive industrial mobilization that occurred as part of the war effort spawned a wide array of technical developments that affected all industrial activity, including forestry. Advances in small-engine design, for example, led to the post-war development of chain saws, which rapidly replaced the use of hand saws and axes in the woods.

Unloading rough lumber at the Williams Lake planing mill, 1950s.

Hydraulics, an entirely new concept in energy transmission, revolutionized the design and function of machines used to harvest and transport timber and the methods of handling logs in lumber and pulp mills. With the war over, and armed forces demobilized, there was a large supply of available workers skilled in the use of these new technologies.

Consequently, areas such as the Cariboo-Chilcotin with its relatively undeveloped forest resource attracted the attention of entrepreneurs in search of opportunity to exercise their talents. Leslie Kerr, with his investment in the small Quesnel mill, was in the forefront of this influx.

The relatively low level of industrial activity in the region in the mid-1940s was primarily due to the character of its forests. The Cariboo-Chilcotin, an area of about eight million hectares,

lies in the rain shadow of the Coast mountains, which provides it with hot, dry summers and, at times, bitterly cold winters.

Along the major rivers—the Fraser and its two principal tributary systems in the region, the Quesnel, which drains the Cariboo in the east, and the Chilcotin, which drains much of the area in the west—the tree species are primarily dry-belt Douglas fir, with lesser populations of spruce, lodgepole pine, birch and aspen. The rolling plateau lands, especially west of the Fraser, are dominated by vast stands of lodgepole pine. Douglas fir and, to a lesser extent, spruce were recognized commercial species and grew to a size utilizable by the milling technology of the day. Lodgepole pine, however, rarely reached a foot in diameter and was regarded as a weed species, useless for lumber production because of its small size and spurned for pulp because of its high resin content.

Before 1950, utilization of these forests was minimal, mostly to supply local markets. The first lumber produced in many areas was sawn by hand, using pit saws. One of BC's first sawmills was built in 1860 at Antler, in the heart of the Cariboo gold fields. Several other small mills—some water powered, and others driven by steam—served the gold rush market and disappeared when it declined.

One of the few mills to survive this era was built by Ithiel Nason at Quesnel to provide timbers for a bridge over the Quesnel River. As Quesnel evolved into the service centre of the region, Nason built a prosperous local lumber business and eventually served the Cariboo as its MLA for two sessions. Another successful Quesnel mill owner of this period, James Reid, served in the Canadian Senate.

A similar sequence of developments occurred at Williams Lake. William Pinchbeck built what was probably the first sawmill in the area, at his ranch. It was likely a water-powered mill, although it may have been steam driven because Pinchbeck had installed a steam boiler at some point in the 1860s to operate a distillery.

A year after the arrival of the railway, in 1920, Harry Curtis built a sawmill to supply the new town with lumber. It was located on the hillside west of town, now known as West Ridge, and its timber obtained by logging the site.

With the evolution of internal combustion engines after World War I, gas-powered motors were readily available to drive small sawmills. By the early 1940s, a dozen or so ranchers in the region had acquired mills to provide for their own needs and, in some cases, to generate a flow of cash by selling rough lumber to their neighbours and residents in nearby settlements.

Leslie's move into the Cariboo-Chilcotin coincided with a major restructuring of BC forest policies. Immediately after the war, the provincial government appointed a Royal Commission under Chief Justice Gordon Sloan to canvas public opinion about proposed changes in the laws governing the use of provincial forests. In parts of the province, primarily the southern Coast and Vancouver Island, and in the Kootenay, forests had been heavily harvested for several decades to supply logs for the well-established forest-product export industry.

On Sloan's recommendation, the government adopted a new set of "sustained yield" policies to limit annual timber harvests, and a new form of lease on Crown forest land, called a Forest Management Licence, to permit the long-term management of designated forests to provide a perpetual timber supply. One of the consequences of these changes was the realization that regional timber supplies were limited. Beginning in the late 1940s, even in areas such as the Cariboo-Chilcotin, where the forest industry was relatively undeveloped, a rush to tie up the available timber supply ensued.

In some parts of the province, the government began granting long-term licences over

Quesnel Sawmills during the 1948 Fraser River flood.

extensive areas of forest to encourage construction of major processing facilities, such as pulp mills. With the exception of a relatively small Forest Management Licence granted to Western Plywood in 1952 to provide a secure log supply for its plants in Vancouver and one opened in Quesnel in 1951, these types of timber allocations were not used in the Cariboo-Chilcotin.

During the late 1940s, the primary means of obtaining Crown timber in the region was through short-term timber sales at public auctions. Typically, these sales were obtained by small sawmill operators. Under policies in place at the time, they were permitted to harvest the larger trees, down to minimum diameter, usually 10 inches (254 mm). Smaller trees were left to reseed the area and grow to provide a future timber supply.

Most of the existing mills in the region were portable and could be moved from one timber sale to the next, producing unfinished lumber. It was either used locally in rough form or trucked to the nearest PGE railway siding and shipped to Vancouver for sale abroad in rough form, or for planing and subsequent sale as finished lumber.

After Leslie entered the partnership with Drezet and Pearson in Quesnel Sawmills, a new operational pattern evolved. In addition to birch logs acquired from Western Plywood, the mill began to obtain fir and spruce logs not useful for plywood.

Within a few months, Leslie purchased Pearson's share in the company and added a planer to the existing sawmill. Then the company began to acquire timber sales, which were contracted out to small sawmill operators who logged them to produce rough lumber that was hauled to the Quesnel planing mill, where it was finished and loaded into boxcars for shipment from Squamish.

Quesnel Sawmills also obtained rough lumber from sawmills holding their own timber sales, either by outright purchase or through more complex financial arrangements. Some of these "independent" operators were financed by Quesnel Sawmills, in which case they were required to deliver their lumber to the company. From Lignum's office at 717 West Pender Street in Vancouver, Leslie sold the lumber, mostly into the growing US market. During its partial first year, Lignum sold $259,575 worth of lumber, earning a modest but encouraging profit.

By the spring of 1948, Leslie was ready to expand the production end of Lignum. After lining up a half-dozen bush mill operators in and around Williams Lake, he secured a 2.8 hectare site along the PGE right-of-way on the northern outskirts of town, across Mackenzie Avenue from the cemetery. One of the operators, Robert Soule, was initially appointed manager of the new business, a planing mill to finish locally cut rough lumber before shipping it to market on the PGE.

The planned new mill had obviously been in the works for some time because, when Soule talked to the *Williams Lake Tribune* about the mill, equipment for it had been stored in town for some time. It was to be equipped with a new Woods 107 planer, a trim saw, a resaw and two sorting chains. A Lodermobile lumber carrier and Hyster lift truck had been acquired to unload and move lumber. Because of the limited power available from the town's diesel generators, the mill planned to install its own diesel power plant.

The mill was built with lumber cut at Quesnel Sawmills and hauled to Williams Lake by one of the Quesnel mill's first employees, truck driver Herb Dell. It was spring, with the frost

*Above: Leslie Kerr (Left) with Bill Kohnke (Centre) and
Frank (Ike) Soule, sawyer, at the Kohnke brothers' Tyee Lake sawmill.*

Right: The Williams Lake mill, looking south on Mackenzie Avenue, 1951.

going out of the Cariboo Highway. Herb spent all day on the road, rarely out of first gear, often bogged down in the middle of the highway. These were standard operating conditions at that time, faced every day by truckers hauling lumber to the planers.

Leslie's initial lumber suppliers included Springhouse Lumber and Soule Brothers, both of whom operated south of town; Jefferson Brothers and Bryce Mills, both from Big Lake; Bert Smith; and Likely Sawmills. On a good day, each of them was capable of producing between 5,000 and 10,000 board feet of lumber. As a result, the planing mill was built to handle 50,000 board feet a day.

The Mackenzie Avenue mill yard, 1951, with Gordon Bruce standing in the office door, behind his truck.

The Williams Lake planing mill between the PGE and Mackenzie
Avenue, with the town cemetery in centre of photo. Note the location
of the rough lumber stored in the residential area awaiting planing.

My father was the least mechanical man in the world. He couldn't screw in a light bulb. And his idea of dressing down was to wear a grey suit rather than a dark blue suit, so he was always quite a figure in Williams Lake. He was never seen without a hat on—not a hard hat but a regular hat. He smoked a pipe, and was very polite, and always bowed if a lady came into the room. A very unlikely character for Williams Lake and Quesnel in the 40's.

He was just a very formal guy compared to anyone in that area. The other guys went to meetings in a combination of coveralls or stuff from the bush and my dad went in his suit from Vienna because that's what you wore to a meeting.

JAKE KERR

This was big news for Williams Lake where, at the time, no business employed more than two or three people. When the mill opened in June 1948, it had a payroll of 15 people. Charlie Cassford, who had come up from Lignum's Vancouver office, had replaced Soule as the first manager. He received $250 a month, the highest paying job in town. Gordon Bruce, who took over as manager when Cassford returned to Vancouver soon after, was also there from the beginning.

One of the first people hired was Roy Astley. He began during the initial construction phase of the mill, which consisted of pouring a concrete foundation for the planer. Once the power supply was hooked up, they began planing lumber. Leslie had orders to fill and there was no time to complete construction of the planer building, which was done later.

From the outset, one inflexible rule, posted in a large notice bearing Leslie's signature, was enforced. Whenever a funeral service was underway across Mackenzie Avenue at the cemetery, all work at the mill had to stop until the funeral ended.

Norman Day, the planing mill's day shift Lodermobile operator, stacking logs at the Springhouse mill.

The new mill produced 4.5 million board feet of lumber at a cost of $44.87 per thousand and sold it for $57.78 a thousand, for a profit of $12.91 a thousand. Sales during the second year of business increased to $383,023, of which almost one-third was on sales of lumber purchased from other mills.

The lumber market was strong, and within the first year at Williams Lake another shift was added. Pete Routley was hired as carpenter to build a maintenance shop—staffed by newly hired mechanic Ken Moore—and an expanded planer building.

By mid-1949, Lignum's mills had become the economic backbone of Quesnel and Williams Lake, especially the latter, which had never had a substantial business of this type. The industrialization of the Cariboo-Chilcotin was underway.

By this time, an operational pattern was apparent. At the bush mills, which were typically set up at the end of a rough road into the middle of a timber sale, a crew of two or three might spend the morning logging. The trees were felled with a hand saw—or a primitive chain saw at an upscale operation—delimbed and topped with an axe. Horses or a farm tractor skidded the logs to the mill.

One Monday morning I went in, and it was colder than hell, winter time. The salesman in Vancouver forgot to tell us about some trucks coming for crossarms. They had to be surfaced. They weren't planed, they were just surfaced. We didn't know anything about it 'til when we went to work that morning and these three trucks were sitting there waiting on us. The truck drivers were told that the lumber was already waiting, so they were mad. They wanted to get on the road and get back to Vancouver. The highway wasn't the way it is today. It was a big trip to Vancouver then.

We were trying to get it through the planer. It was frozen, and we had a hell of a time. Well, about two in the afternoon, we had one truck loaded and half of the other two, and Leslie came up the yard. There was a funeral going on on the other side of Mackenzie Avenue. Jesus, did he blow his steam!

He called me and the planerman in—I was a grader—and he figured it was our responsibility. He had us standing up against the wall and was shaking his finger right under our noses. He had a notice up on the wall there that everything must come to a stop in the event of a funeral across the street. I'll remember that day as long as I live.

Roy Astley

The afternoon was devoted to sawing the logs into lumber, with one person on the head-saw, another on the edger and the third one removing slabs and piling boards on the deck of a three-ton truck. When it was loaded well beyond its rated capacity, with three to four thousand board feet, a chain was thrown over the lumber and cinched down. The crew crowded into the cab and headed for the planer mill in Quesnel or Williams Lake, usually through clouds of dust, miles of mud, or drifting snow.

Wilbur Ure delivering a load of rough lumber to the Mackenzie Avenue mill in the 1950s. (L to R) Lignum employees Carl Thiede, Jin Abe and Andy Anderson.

At the mill they got in line with other trucks just like them, waiting their turn for unloading. At Williams Lake, Larry Toews was the first fork lift operator to unload trucks. He and his dad, Cornelius, were the first in a long line of Lignum father-and-son employees. Cornelius was a lumber piler and one of the first employees to retire at Lignum, after nine years of work. Larry stayed for 20 years.

When business was brisk and the roads passable, piles of lumber were stacked along Mackenzie Avenue, clogging up the mill yard. When a truck was unloaded, Gordon Bruce scribbled out a tally on a scrap of paper and stuffed it in one of the many voluminous pockets of his bush jacket. Apart from that, his records consisted of three large spikes on the office wall.

Gordon was a man of direct action, backed up by a wicked sense of humour. On one occasion, when a fire started in the office washroom, he leapt onto a forklift and tore the burning washroom away from the office building. Another time, he decided the greenchain crew was lingering longer than they should in a small, heated shack where they ate lunch. Gordon took a section of a broom handle wrapped in brown paper to resemble a stick of dynamite and attached a short length of fuse.

One day after lunch he kicked open the door of the lunch shack, lit the fuse and tossed the stick into the middle of the room. The lunch room emptied in seconds. Jim Deutch, a big man standing about 6'4", went through a window much smaller than himself without slowing down. That ended long lunch breaks for the greenchain crew.

Gordon Bruce, general manager, at the planing mill sorting chain, 1950s.

The quality of the rough lumber delivered to the planer varied. Some mill operators were good sawyers and knew what they were doing. Their boards were of more or less uniform thickness and standard widths. Others had little experience and their lumber could be an inch thick at one end and two inches or more at the other. Of course, they were convinced they had delivered a load of two-inch lumber and sometimes were appalled and outraged to find otherwise after it had been planed. Part of Gordon's skill as a manager was to calm them down, show them how to do a better job and convince them to keep hauling in the lumber. In some cases he succeeded, in others not.

But the markets kept growing, Leslie always had more orders to fill than lumber to ship and the number of bush mills operating continued to increase. By 1950, the Quesnel mill was operating at full capacity and the one at Williams Lake was unable to obtain enough rough lumber.

That year, the Quesnel mill burned. Leslie bought out his remaining partner in the operation, Fred Drezet, and built a new planing mill at Two-Mile Flat, at the end of the PGE line.

By then there were half a dozen planers operating around Quesnel and the competition for logs was getting stiff. At the same time, there was a good market for rough lumber in Europe, which provided even more competition. Lignum responded in various ways.

Victor Brown-John's bush mill at Likely was purchased and put on two shifts. Charlie Cassford came up from Vancouver to run the logging end of the operation. Lignum began buying timber sales in the Likely area and put a second mill into operation, under an operating contract with Arthur Goyette.

Around this time, Lignum initiated a practice that underlay its future success. Most bush mill operators could not obtain financing for equipment purchases, timber sale deposits, or to accumulate log inventories during slack markets. Since Lignum's business was dependent on their lumber supply, Lignum began financing these operators through outright loans or advances on lumber production. In exchange, operators were required to deliver their lumber to a Lignum mill. These were known as "controlled" mills.

Some of the bush mills were independent, at least some of the time. They sold lumber to whichever planer offered the best price on any given day. In the early stages of the region's development it was not uncommon for an independent operator to cut a load of lumber and head for town to shop it around the mills.

In the highly competitive Quesnel area this practice sometimes undermined longer-term arrangements of the sort Lignum introduced. When the bush mill operators started moving into fir stands west of the Fraser River, they all had to bring their lumber in over the area's only bridge. Planer operators with a pressing need for rough lumber would park along the road west of the river, their pockets full of cash, waiting for an independent bush mill operator to come along with a load to sell. It was not unknown for controlled operators delivering a load to make a quick cash sale, hoping the transaction would go unnoticed by the mill financing them.

In the early, frontier era of the Cariboo lumber industry, a mill manager needed to keep a sharp eye on everything. Later on in the 1950s, for instance, it became apparent something was amiss at Lignum's Likely mill. Each month, 10 to 15,000 board feet of rough lumber shipped from the mill failed to reach the Williams Lake planer. A crew was dispatched to Likely to repile the entire lumber inventory and discreetly mark the end of each board with an inconspicuous dab of paint.

In the weeks that followed, Gordon Bruce made it a point to drop in at competitors' planing mills for an informal visit. During one of these visits he watched two slings of lumber bearing the paint marks being fed into the planer. The next day a thorough shakeup of truckers hauling out of Likely occurred.

Top: Lignum's Likely mill, April 1951.

Bottom: Bunkhouses at the Likely mill, 1951.

In 1951, two larger mills started up in the Williams Lake area. They were a notch up the scale from bush mills, in a fixed location and with nearby, long-term timber supplies. One was built at Dugan Lake, northeast of town, by the three Kohnke brothers, Walter, Felix and Bill. They were professional wrestlers who had operated a sawmill in the Fraser Valley before coming to the Interior.

Fred Linde and his brothers built the second mill near Springhouse, south of town. They had grown up on a sawmill operation in South Dakota and after military service during the war and a stint in Oregon had gravitated to the Cariboo.

Top: The Kohnke Brothers' Dugan Lake mill crew, with Bill Kohnke in the middle.

Bottom: The Kohnke Brothers' Dugan Lake mill, April 1951.

Both the Kohnkes and the Lindes were experienced sawmillers and quickly became major suppliers to Lignum's Williams Lake planer. The Kohnkes operated in an exclusive association with Lignum for many years, eventually selling their operation to Lignum. The Linde mill evolved into an independent company with its own planer, which is still in operation, and has sold a portion of its lumber production to Lignum for more than 50 years. These two mills, along with Lignum's two mills at Likely, supplied most of the Williams Lake planer's lumber for many years.

By 1951, there were more than 100 people employed in the Lignum-owned mills, and production had increased to 14 million board feet. That year, they were certified by the Prince George local of the International Woodworkers of America. It was the first company organized by the union in the region.

Unlike many of his colleagues in the industry, Leslie had no quarrel with unions. He was astute enough, however, to offset the Lignum contract by a year from the Prince George contracts, insuring any strikes there would be settled before Lignum's contract expired.

As production at the two planing mills increased and the North American lumber market picked up speed, the sales office in Vancouver also grew busier. It was a lean operation at that time, consisting of Leslie, who spent a good bit of his time travelling back and forth between Quesnel and Vancouver on the PGE, Fred Sleep, who looked after sales, and Jerry MacLurg, who served as Leslie's assistant, office manager, purchasing agent and in any other capacity not otherwise covered. Lignum was beginning to sell lumber for other mills in the Interior, and between Sleep and Leslie they made most of their early sales in Europe, slowly easing into the US market as they gained momentum.

In 1951, a reorganization of the sales department began when Leslie brought a cousin who had recently immigrated from Europe, Emil Turner, into the company. Lacking experience in the industry, he was dispatched to the Quesnel mill to learn the lumber business from the bottom up. After a year or so he returned to Vancouver, where Leslie had set up a separate trading company, Lumbox Trading, incorporated in January, 1952. Six months later its name was changed to Lignum Sales Ltd. Emil eventually became sales manager in this division.

This established Lignum's basic operating structure—a production division in the Cariboo-Chilcotin and a marketing division in Vancouver—which would survive for the next 50 years.

The 1974 Lignum Reunion included many of the Williams Lake mill's first employees. (L to R)
Rear Row: Dick Wainwright, Stuart Smith, George Abe, Tim Kerr, Roy Astley, Franke Thiede, Tom Bishop, Ken Moore,
Charlie McQue, Art Goyette
Third Row: Max Butler, Jin Abe, Walter Kohnke, Sam Abe, Gary Ingals, Corny Toews, Herb Dell, Wendell McDonald
Second Row: Joe Benze, Dane Powell, Jammi Abe, Cliff MacIntosh, John Ailport, Vic Siedan, Gordon Bruce, Joe Bott
Front Row: Pete Routley, Walter Ivans, Wally Langford, Larry Toews, Jake Hooge

CHAPTER 3

THE MOVE TO WILLIAMS LAKE

Lignum's early successes—establishment of a viable planer mill at Quesnel, the expansion to Williams Lake, and the increased production of both mills—did not go unnoticed. Within a few years, several other operations established themselves in the region, patterned loosely on the Lignum model—a central planing mill supplied by a number of mobile bush mills.

The first of these new ventures was built at Horsefly Lake, about 75 kilometres east of Williams Lake, by the Gardner brothers, sons of a long-established rancher and bush-mill operator near Wells, in the foothills of the Cariboo Mountains. In 1928, Jack Gardner launched a retail building supply outlet in Quesnel that included a small planer to finish lumber cut at his ranch. Gardner's company supplied local needs and most of the bridge timbers for the PGE.

In 1951, his sons opened the Horsefly operation, a combined sawmill, planer and moulding plant, with a capacity of 40,000 board feet a day. They also opened a retail outlet in Williams Lake. Their combined operations employed 45 people. That year, Western Plywood opened a veneer plant at Quesnel.

Several other operations were also established around Quesnel at this time. John Ernst came from Nova Scotia to open a small mill. The Brownmiller brothers, who arrived from northern Saskatchewan, owned two mills, along with a share in a third started by Cliff Falloon, all three of which were eventually sold to West Fraser. Another small, short-lived planing mill was operated in Williams Lake by the Argus Lumber Company. Yet another was started by Joe Faulkener, who sold out to Bert Siemen, who went belly up. Many similar operations came and went throughout the region, leaving little trace of their existence apart from a pile of sawdust and, in some cases, a trail of unpaid bills.

Two new mills opened in Williams Lake in 1953. All-Fir Lumber, with the first electric-powered planer in the region, was built by three American partners. Five years later, the US-based Merrill family—represented by Corydon Wagner—invested in the mill and later that year it merged with the Gardner family firm, and renamed Merrill & Gardner. In 1967, the name changed again, to Merrill & Wagner. In 1977, it was sold to Weldwood, the successor company of Western Plywood.

Kohnke Brothers truck hauling fir to their Tyee Lake Mill. Truck equipped with side loader.

The second new mill was established by Gabe Pinette and his in-laws, Dollard and Roger Therrien. Originally loggers and bush mill operators from northern Ontario, they had been milling timbers on the British Properties in West Vancouver. When that operation collapsed, they got their hands on a portable mill and headed for the Interior. Running low on money and gas by the time they reached Williams Lake, they settled in and began cutting rough lumber, which they sold to Lignum, Faulkener and Siemen. When the latter went broke, Pinette and the Therriens built their own planer, along with a new stationary sawmill, the first sawmill of this era built in Williams Lake. They quickly became Lignum's first real competitor in town.

A year later, in 1954, another serious contender appeared when Harold and Odt Jacobson bought a burned-out bush mill with a timber sale at Beaver Valley, near Horsefly. The Jacobsons had grown up working on their father's water-powered mill at Terrace, and worked logging in several parts of the province. They rebuilt the mill and added a planer, relocating the entire operation to Williams Lake 10 years later.

The last of the pioneer Williams Lake mills was not built until 1957, when the Ketcham brothers opened a cant and planer mill in the Glendale area. The brothers—Bill, Pete and Sam—were from Seattle, where they had grown up working in their father's wholesale lumber company. In 1955, the boys struck out on their own, buying the small Two Mile Planing Mills in Quesnel. Sam settled in the Cariboo to run the production end of the

> *At one time I had 22 horses, one man per horse, logging. They would log it and bring logs to the mill. We'd buy it by the truck-load. All these small logs with 6" tops. I couldn't get enough scalers to keep going.*
>
> *So what we did, we started by measuring the load, the sizes of it, and working out the cubic measure. That works out if people are honest enough. But the first thing you know, the logs were piled like straw, they weren't piled properly. They come in with a big load and there was nothing in it.*
>
> *So we started weight scaling, and the Forestry went along with us. We had our own scales. All the mills today have their own scales. We were one of the first to do our own weight scaling.*
>
> GABE PINETTE

A typical Cariboo lumber truck of the 1950s, hauling a heavy load of rough lumber from a bush mill over a poor road to a planing mill.

operation, acquiring a string of bush mills on both sides of the Fraser. His brothers handled the sales end in Seattle. Calling their company West Fraser Timber, they pioneered the process of cutting cants in the bush, and milling them into boards at stationary scrag mill and planer operations such as the one they built at Williams Lake.

The same general process was occurring throughout the Cariboo. By the mid-1950s there were probably close to 500 bush mills operating in the region—one behind every stump, as the saying went. From Quesnel south along the PGE, in virtually every centre, planers were built to finish the rough lumber cut by the bush mills.

The Ainsworth brothers, former falling contractors from Vancouver Island, bought a portable mill mounted on wheels and towed it to 100 Mile House where they cut rough lumber for Cariboo Lumber. The latter was started by Leslie's old friend from his days in Paris, Laci Kardos. He immigrated to Canada in 1951, and Leslie took him up to the Cariboo for a look around. His operation evolved into Kardos Forest Products, which he sold in 1962 to Komori Lumber.

Lignum's Vancouver sales office sold lumber for several of these mills, some when they were beginning and others, such as Kardos', throughout their operating lives.

The Ainsworth brothers' first portable mill enroute to the Cariboo, 1952.

This rapid expansion of the Cariboo lumber industry during its first decade of full-scale operation was, by 1955, creating regional timber shortages. The amount of timber available in an area at a given time was, to a large extent, a matter of definition. Two important considerations were the species and size of trees utilized.

At the time Lignum came to the Cariboo, the species most commonly cut for lumber was Douglas fir. There was an established market for it, there was a supply available near settled areas and transportation corridors, and lumber milled from it could be shipped across the continent in an undried form without a deterioration in quality. Only Douglas fir trees more than 12 inches in diameter were harvested, and their tops were cut off at eight inches diameter.

Leaving smaller fir trees was a conservation measure designed to ensure the growth of the next timber crop, but the generous top allowance was one example of the enormous amount of waste the industry generated at this stage in its development. The bush mills created even more waste in the form of slabs, poorly cut boards and sawdust. Every mill generated a huge pile of wasted timber, often accounting for more than half the volume harvested.

Forests in the region contained substantial volumes of other species, particularly spruce and lodgepole pine. Some spruce was milled for local use, but relatively little was shipped because it required drying to avoid discolouration and decay during shipping. Spruce sold out of the region was normally air dried at the planer before shipping. Green spruce loaded in a rail car and shipped to, say, Miami, would be mouldy when it arrived.

Lodgepole pine also required drying to avoid discolouration during shipping, but it was not easily air dried. A further deterrent to the use of pine was its small size. Lodgepole pine in the Cariboo-Chilcotin was mostly found in pure stands of trees less than a foot in diameter. In some mixed stands along the rivers it occasionally grew to larger dimensions, useable by the mills of that era and sold as spruce. A lot of pine trees were not very straight, and generally it was considered a weed.

When the issue of timber supplies arose in the region in the mid-1950s, the shortages

experienced were primarily Douglas fir and—to a lesser extent spruce—more than 12 inches in diameter. There was no shortage of mature lodgepole pine under 12 inches in diameter. In fact, most of the country west of the Fraser River, and much of the region east of the river, was carpeted with it.

Much of the Cariboo-Chilcotin region is carpeted with lodgepole pine.

A second factor that came into play at this time was the application of sustained yield policies adopted in the wake of the first Sloan Royal Commission. This included establishment of sustained yield management units and the calculation of an "allowable annual cut" (AAC) for each unit. Until this task was completed, timber was sold to the highest bidder in an unrestricted manner.

As the number of mills in the Cariboo increased, so did competition for the timber. By 1953, public auctions for timber in the Williams Lake sustained yield unit were attracting crowds of mill owners, and by 1955 the winning bids were often too high to permit profitable operations.

At about this time, the Forest Service calculated the AAC for the Williams Lake unit to be 35 million board feet. When the district forester, Lorne Swannell, announced this figure at a meeting of 85 alarmed lumbermen in Williams Lake—the largest such gathering ever—he pointed out that the previous year, 65 million board feet had been logged in the unit. Although the Quesnel Lake district harvest was less than its AAC, the 100 Mile House district was in even worse shape than Williams Lake.

(The AAC and the actual harvest figures related primarily to Douglas fir, with some consideration for spruce and large pine logs. Lodgepole pine under 12 inches in diameter was not included in the AAC; nor was it harvested and milled into lumber.)

A contributing factor to the growing concern about the timber supply derived from another aspect of the government's new policies, Forest Management Licences. These were another form of sustained yield tenure, managed by private forest companies. If an application for an FML was accepted, a reserve on the timber within it was granted and was not available to mill operators in the form of Timber Sales. To date, only one FML had been granted in the region, to Western

Plywood, but applications for four others, including one from Lignum, had been submitted to the Forest Service. Reserves had been placed on these areas excluding local bush mill operators and others.

From its inception, this new licence was extremely controversial. Its very creation had been opposed by much of the established forest industry, especially on the coast. Leading figures in the industry, including H.R. MacMillan, said it would drive out small independent operators, lead to monopoly control of the timber supply, and fail to provide the required incentives for the private sector to invest in the management of future timber crops.

The Coalition government of the day ignored these arguments and when the large established coastal companies boycotted the licences, the government began granting them in very large blocks to big out-of-province forest companies with no previous operations in BC. The government's policy appeared to be one of granting exclusive, perpetual licences over large areas of Crown land in exchange for the building of costly manufacturing plants, such as pulp mills.

For instance, the first FML granted was to Celanese Corporation, a New York company, for several million acres of prime timber land in the Nass and Skeena valleys. Coincidentally, the licence was located in Coalition forest minister Edward Kenney's riding. The deal included construction of a pulp mill in Prince Rupert—an ill-fated operation that haunted provincial governments and undermined the forest economy in the northwest corner of the province for more than 50 years.

Although a few smaller FMLs were granted to established Interior companies, such as Western Plywood, uncertainty about the future of the timber supply was rife. The anxiety soon became mixed with anger when rumours began to circulate that FMLs could be bought with bribes.

Two other events during this tumultuous period affected the situation. In 1952, the Social Credit government under WAC Bennett was elected, in large part because of its promise to foster industrial development in the province, especially in the northern Interior. One strategy adopted by the Bennett government was to attract outside capital with the generous allocation of Crown resources, including timber. It was Bennett's forest minister, Robert Sommers, who was suspected—and in 1958 convicted—of accepting a bribe to grant a FML to BC Forest Products Ltd., a large new corporation created by Toronto financier E.P. Taylor, who had no previous business interests in BC.

The second event was the extension of the PGE from Quesnel to Prince George, also in 1952. Because of the connection it established with the Canadian National Railway's transcontinental line, the extension paved the way for Cariboo lumber to enter the prairie and eastern Canadian markets, and offered a shorter route to Asian export markets via the Port of Prince Rupert. It also provided a connection between the Cariboo and Prince George forest industries that would soon become crucial to them both.

Lignum's application for a FML was submitted in December, 1954, and in response to a request from the Forest Service the following June a submission was prepared, which included a detailed description of the company's operations. An accompanying letter from Leslie stated:

> *The Lignum group of companies operates in the Williams Lake and Quesnel districts. Complete lumber manufacturing facilities are maintained at both places. Investments to date in plants and equipment amount to about $700,000. Loans to operators and timber sale deposits bring the total investment to approximately $1,000,000. More than 50 contractors and partially financed operators are associated with the Lignum group. These associates and the Lignum group, together have about 450 employees.*

Top: *The sorting chain crew at the Mackenzie Avenue planing mill in the 1950s.*
(L to R) *Corny Toews, Vic Carlson, Clarence Moore, Wolf Oerfelt, Karl Thiede, Percy McKee, Dick DeGues, Walter Caspard*

Bottom: *Ken Moore with the Lodermobile at the Mackenzie Avenue mill, 1951.*

By the mid-1950s both Lignum and Williams Lake were growing, with lumber stacked wherever space could be found, on both sides of the highway and the PGE mainline to Squamish.

Given the volatile nature of the province's forest economy during the period since Leslie first invested in Cariboo Birch Mills in 1946, the state of the company was a notable achievement. Leslie had parlayed the few thousand dollars he was able to raise on the family house in Vancouver into a million dollars in assets, financed mostly from profits earned by the production divisions in Quesnel and Williams Lake, and the sales division in Vancouver. It was, at this point, the largest and most stable economic endeavour in the Cariboo.

Lignum's long-term stability depended on an unsure, short-term timber supply, as indicated in Leslie's letter:

> *Total lumber production for 1954 was 60,000 MBM. For 1955 and annually thereafter it will be about 75,000 MBM. Raw material requirements are 55,000 MBM of timber annually. The estimated volume of timber remaining on timber sales held by the Lignum group as at June 15, 1955, is approximately 97,000 MBM or less than enough for two years operations. The raw material supply situation for the Quesnel Division is particularly serious.*

The description of operations revealed the workings of a unique set of relationships between Lignum and its associated contractors, semi-independent and independent operators. Lignum itself consisted of three divisions. The Williams Lake Division included the planer in town, the sawmill at Likely, and Cariboo Fir, a small logging operation with timber sales of its own. Thirty-three bush mills were financed or under contract to the division.

The Quesnel Division consisted of the planing mill at Quesnel, sawmills at Quesnel and Cinema (north of Quesnel), and Hush Lake Spruce Mills, which owned a sawmill east of Cinema. Twenty-five bush mills were associated with the division. The Vancouver Division comprised the Lignum head office and Lignum Sales.

Lignum's Likely sawmill c. 1955, with rough lumber awaiting transport to the planer at Williams Lake.

The output and operation of the Quesnel and Williams Lake divisions were identical. They produced finished lumber from rough lumber obtained in a variety of ways, including a portion obtained from each division's own sawmills.

Another portion was received from contractors or semi-independent operators running bush mills on timber sales owned by the divisions. They were required to deliver all the lumber they cut to Lignum planers. In effect, they were all contractors. This "controlled" group included bush mill operators holding timber sales of their own who were financed by Lignum. They too were required to deliver all their lumber to Lignum planers.

The final source was rough lumber purchased from independent operators with no financial ties or production obligations to Lignum.

In 1955, the breakdown of these supply arrangements was:

Source	Williams Lake Division MBM	%	Quesnel Division MBM	%
Division sawmills	7,000	20	6,000	15
Controlled sawmills	21,000	60	14,000	35
Independent operators	7,000	20	20,000	50
Total	35,000	100	40,000	100

Half the lumber obtained from independent operators at Quesnel came from a gang mill operated by Western Plywood at its local veneer plant, and was finished under contract to Western Plywood, an agreement that would terminate at the end of the year.

Lignum Sales sold all the finished lumber produced by the two divisions, except for the volume planed for Western Plywood, as well as lumber from other Interior mills.

At about this time, the Forest Service began implementing an informal system of timber allocation which had no basis in law but was accepted by industry because it put an end to the often-chaotic public auctions. Operators within public sustained yield units where the AAC was fully allocated were given preferential bidding procedures on timber sales. This procedure evolved into a "quota" system in which licensees were more or less guaranteed a fixed volume of timber, equivalent to their historic allocation. In time, "quota" became a saleable commodity and assumed a value of its own.

Top: *Ken Moore on the forklift unloading rough lumber from Wilbur Ure's truck at the Mackenzie Avenue mill.*

Bottom: *Rick Toews, a sorter, at the head end of the Mackenzie Avenue mill's sorting chain, 1952.*

In 1955, the government convened a second Royal Commission under Chief Justice Sloan. Although the commission was given a broad mandate, it soon became focussed on the contentious issue of FMLs. By this time, the Forest Service had received more than 100 applications for FMLs, including Lignum's, and the government had put a freeze on their processing.

At a commission hearing in Quesnel, a brief presented on behalf of Western Plywood suggested that, because small operators would inevitably disappear, the government should favour larger companies that could provide future public revenues and finance large industrial plants to give secure employment to local communities.

This argument was vigorously countered at a Williams Lake hearing in a brief presented by F.G. Butler on behalf of the Horsefly Committee of the Williams Lake and District Board of Trade. He argued that if multiple use of the forests and community stability were objectives, then the existing operators should receive preference over large outside corporations. The battle lines were drawn.

During a commission hearing in Vancouver in December, 1955, a lawyer appeared with evidence of bribes given to the forest minister to obtain a FML. Sloan refused to accept the information, and the lawyer turned it over to the press. Pandemonium ensued, and had abated only slightly when commission hearings resumed in January in Victoria. Leslie, accompanied by Lignum's lawyer, Edgar Grossman, was one of the first witnesses to appear.

Grossman was a young lawyer who had helped Leslie with his legal affairs from the time Leslie left Alaska Pine. He was involved in setting up the Garibaldi mill, and in its demise. And he had handled the incorporation of Lignum and its associated operations. By 1956, he was providing Lignum with more than legal advice and had become a key advisor in the formulation of the company's strategic plans. A half century later, his law firm, Grossman and Stanley, still provides Lignum's legal services and Grossman, from semi-reclusive retirement where his interests range from theoretical physics to foreign languages, still plays an active role in planning the company's overall business strategy.

In their appearance before the commission, Leslie and Grossman worked as a tag team, with the lawyer asking questions to illuminate and elaborate on the Lignum brief, which was read into the record by Leslie. The joint presentation elicited a flurry of questions from the commission's counsel, prompting Leslie to outline a comprehensive approach for development of the industry in the Cariboo.

The formal brief outlined Lignum's application, which was for two blocks of land. The first, 76,950 hectares in size, was near Likely and the second, comprising 182,250 hectares, was southeast of Williams Lake. The two blocks had a combined AAC of about 2.4 million board feet.

The brief also described Lignum's forest management vision:

Timber is a crop to be sown and cultivated to provide an annual harvest. The annual harvest is not to be limited to the natural regrowth of the species harvested. Change is one of the few certain attributes of human endeavour. In place of natural regrowth men can develop, sow and cultivate more desirable species and can devise means of encouraging faster growth. Material only recently considered waste can now be utilized. It will become possible to manufacture materials of economic value from trees and parts of trees not now merchantable...

It continued with a discussion of harvest rates of the forests managed under such an approach. In view of the depredations of the Cariboo forests by insects since the brief was written, its proposals were noteworthy:

> *Timber currently merchantable should not be cut at its rate of natural regrowth. The rate of cut should be increased to encourage faster growth of young trees and to clear land for new and improved cultivation and to prevent loss by deterioration and to make allowance for new methods of production and utilization.*

Under Grossman's questioning, Leslie explained his proposals, with reference to his European experience:

> *I feel that the present concept of sustained yield is to begin with a static concept and there is nothing static can exist, and it is unrealistic because it doesn't take into account future developments. Improved methods of logging and hauling will be used in the near future, improved methods of conversion and waste utilization will prevail very soon. Different species might be more desirable in the future. Different products might be made and also forestry techniques and thinking regarding crop rotation are likely to change. Now, to me, the sustained yield concept based on natural reproduction is something like the Maginot Line . . . I lived in France and that Maginot Line didn't pan out... it did when it was thought of. . .but their mistake was that they just took the thinking of the period and disregarded possible revolutionary changes. With the changes we have seen in the last ten years, I just cannot imagine that any policy formulated on a long-term basis should be based on the natural rate of reproduction. I certainly do not advocate that no attention should be paid at all to the rate of reproduction, but as a long-term policy, certainly considerable allowance should be made for changes.*

Having established Lignum's proposed approach to managing forests, Grossman raised the then explosive question of conflict over timber rights and the growing dispute between small and large forest companies. His line of enquiry was taken over and pursued in detail by the commission's counsel.

During the subsequent discussion Leslie described a unique and, to those looking for a resolution to the growing conflict between small established firms and large outside ones, intriguing set of policies. Almost 40 years later his ideas were described in a perceptive graduate history thesis about Lignum written by Mary McRoberts at the University of Victoria.

When Leslie Kerr presented his brief to the Victoria sitting of the Sloan Commission in early 1956, it was clear he intended to enhance the district (forest industry's) structure through a long-term development plan, rather than simply replace existing enterprise with a totally new form of lumber business. In Kerr's view, the existing structure held the key to the district's future and, as such, needed the protection of long-term tenure to expand and grow into an increasingly efficient economic sector. Kerr agreed that the economies of scale required to carry out the extensive forest management requirements of the FML were often too great for smaller operators and contended that they were unfair to most district operators by denying them access to any form of long-term tenure. As a possible solution, Kerr recommended that the Forest Service grant a small operator long-term tenure if he was able to get a large forest company like Lignum to guarantee the forest management requirements. The small operator would retain the timber in his own name, and the tenure was only valid if that operator remained in the business. The guarantor benefited from the lumber sold to him for finishing and/or shipment, but was forced to maintain competitive prices in order to ensure his supply source remained an economic venture.

Generally, over time, Kerr maintained that consolidation and centralization would occur as entrepreneurs reinvested profits and upgraded their plants, thereby making opportunistic ventures increasingly inefficient. In Kerr's long-term plan, consolidation and growth would occur from within the sector rather than through imported capital. The government's goal of increased forest revenues would be achieved through a gradual increase in the "average efficient operation" level upon which minimum stumpage was based and by the industry taking over the forest management role of developing the district timberland and improving the wood value of the forest cover. By putting the onus on industry to ensure that standards of operation constantly improved, and by rewarding efficient operators with a guaranteed timber supply, community development through industrial stability and growth could occur while contributing greater revenues to the provincial government through both stumpage and corporate tax dollars.

Leslie's ideas were consistent with those of other critics of government policies and of other proposals to achieve the stated objective of developing and stabilizing local economies. Unlike the others, though, Leslie's ideas would, to a large extent, be realized.

During his testimony, Leslie also described Lignum's future development plans. While discussing the increasing distances rough lumber was being trucked to have it planed and shipped, he indicated an intention to centralize Lignum's operations in Quesnel.

We are hauling lumber from Likely to Williams Lake, almost 70 miles, which is a temporary feature because we expect to take the whole Quesnel Lake production to Quesnel, once the Lake Quesnel road is fixed, which will cut the haul down to about forty or forty-five miles from the present seventy-five, plus transferring the lumber from Williams Lake to Quesnel by PGE.

If this, indeed, was the intention it may have been adopted in view of the timber supply available in the two districts. Block 1 of Lignum's FML application, located in the Quesnel district, was only two-fifths the size of Block 2, which lay in the Williams Lake district. But Block 1 contained more than twice the volume of merchantable timber as Block 2.

Block 1, of course, contained substantial volumes of Douglas fir and spruce more than 12 inches in diameter, timber that was merchantable by the standards then in effect and for which Lignum's operations were built. Block 2 contained higher volumes of lodgepole pine which, by Lignum's and most others' standards, was not merchantable.

**Vic Bremmer Trucking
hauling logs for Dean Getz
to a bush mill, 1958.**

Whatever the plans were, further development of the Quesnel division was soon abandoned. The government did not buy into Leslie's proposals for a FML and, although it continued to issue licences in other parts of the province for another five years, no more were issued in the Cariboo-Chilcotin region. The possibility of improving the Quesnel division's timber supply began to fade.

Also, by late 1956, the North American economy began to slide into a mild recession. Lignum's mills continued to operate as the company fought to ride out the downturn. By late 1957, things were much worse. The market had taken another plunge in the early fall. Leslie traded in his Cadillac for a Chevrolet to help keep the bankers off his back. It snowed early and deep that winter, and temperatures dropped to unusually low levels. Many bush mill operators were wiped out.

In mid-winter the larger operators met to form the Cariboo-PGE Lumber Manufacturers' Association to address problems common to all lumber producers operating along the railway. It

included firms in the Kamloops and Prince George forest districts, which at that time included Williams Lake and Quesnel. Leslie was elected chairman. Objectives of the association were to obtain equitable freight rates from the PGE, negotiate increases in the allowable annual cut, provide commentary on the Sloan Commission recommendations, and establish safety and lumber grading programs.

For the first time since the company was formed, Lignum had difficulty meeting its payrolls. The cheques were delivered as usual, but with a note attached asking that they not be cashed until further notice, which often was only a day or two, but sometimes as much as a week.

Just when things looked like they could not get any worse, disaster struck. Western Plywood terminated its supply of rough lumber to the Quesnel planer. Since this source provided 25 percent of the entire Quesnel division's supply, drastic measures were required.

Leslie met with John Bene and worked out a deal to salvage a viable portion of Lignum. Western Plywood would advance funds to Lignum on a scheduled purchase of Lignum's timber

Above: The crowded Mackenzie Avenue mill yard in 1955, with forklift carrying rough lumber to planer.

Left: The Mackenzie Avenue mill in 1955, with rough lumber being fed into planer. Pipe at upper right carries waste to beehive burner across the PGE mainline.

rights in the Quesnel division two years hence. Lignum would refrain from acquiring additional timber rights in the Quesnel area, and Western Plywood would begin a withdrawal from the Williams Lake area. Lignum's mills in the Quesnel division and at Likely were either closed or sold, some of them to the Brownmiller brothers.

Included in the deal was $750,000 cash, which Lignum used to clear its debt and expand its operations in the Williams Lake division. Its first move was to buy a sawmill with a good timber supply from Scotty Frizzi at Springhouse, south of Williams Lake. Additions were made to the mill to increase production to 30,000 board feet a day, and a planer installed. John Tresierra was dispatched from Williams Lake to oversee the Springhouse planer, and Bob Elliott was placed in charge of the sawmill.

Scotty Frizzi's Springhouse sawmill in 1950, which Lignum purchased in 1957.

Another purchase was made in the Horsefly area with the acquisition of OK Sawmills, including substantial timber rights. A bush mill operator in the Beaver Valley area, Albert Steckle, had recently obtained timber for a mill he intended to build, but instead sold it to Lignum. Some of the bush mill operators who possessed their own timber sales and had gone broke the previous winter were indebted to Lignum. Their timber positions were also taken up by the company, along with those of other operators who wanted to sell their rights and get out of the business.

The reorganization of Lignum's production division in Williams Lake coincided with a general improvement in the lumber markets. By mid-1959 the mills were running flat out to fill the orders pouring into the Vancouver sales office. Then, disaster struck again.

One September Sunday night a fire started in the Mackenzie Avenue planer mill, burning it to the ground. The crew worked day and night rebuilding a temporary replacement to finish the rough lumber pouring in from the bush mills, and to fill the mounting pile of orders.

The site was already overcrowded, with lumber carriers running up and down the Cariboo Highway or back and forth across the PGE mainline to store lumber on rented land. There was no room for any expansion of the planer facility, let alone the addition of a sawmill.

The solution was to move to a new site, along the railway on the other side of town. Land was leased, a sawmill built and, on Labour Day weekend, 1961, the planer was moved to the new site.

Lignum's production division had found its permanent home.

The aftermath of the fire at the Mackenzie Avenue mill in 1959.

CHAPTER 4

A DIFFICULT DECADE

DURING THE DECADE OF THE 1960S, LIGNUM, ALONG WITH THE REST OF BC'S FOREST INDUSTRY, UNDERWENT A PRODIGIOUS TRANSFORMATION. SHEER SURVIVAL THROUGH THIS PERIOD WAS A SIGNIFICANT ACCOMPLISHMENT. THE WEEDING OUT OF FOREST COMPANIES IN THE CARIBOO WAS RELENTLESS. THOSE REMAINING IN BUSINESS AT THE END OF THE DECADE DID SO MORE BECAUSE OF THEIR FLEXIBILITY AND ABILITY TO ADAPT TO NEW CONDITIONS THAN BECAUSE THEY HAD WELL-THOUGHT-OUT LONG-TERM PLANS.

By the same token, the skills and capabilities needed to survive the 1960s were far different than those needed to thrive during the subsequent decade.

Throughout most of the 1960s, forest-product markets were strong. Demand for dimensional lumber in the US grew steadily.

The most pressing problem facing Lignum and its competitors in Williams Lake was timber supply. Williams Lake was farther away from the best Cariboo Douglas fir sites, giving Quesnel mills an advantage in that market. The untapped lodgepole pine stands of the Chilcotin plateau west of the Fraser River beckoned, but most of these trees were too small to handle efficiently with the existing logging and milling technology.

Lignum was relatively well supplied with fir to feed its mills, at least in the short run, so the pressure was not so severe as it was on some of the other operators. Pinette and Therrien—like West Fraser, and the Ainsworths at 100 Mile House— did not enjoy that luxury. They had to get into pine or get out of the lumber business in the Cariboo. So they were the first to switch species.

The conversion of the Cariboo forest industry was facilitated by, perhaps, the most intense period of technological change in the history of the industry. In the logging end of the business, most of these developments occurred elsewhere and were brought to the Cariboo fully formed and intact. Innovations in milling lumber, however, were developed with the Cariboo lodgepole pine forests in mind, and revolutionized lumber production around the world.

At the beginning of the 1960s, logging in the region was a fairly labour-intensive business. Trees were still felled by hand, using power saws that had become available during the previous decade. These saws were also used to delimb and buck logs to length at the stump. Horses and, increasingly, farm tractors or

The first mill, we'd designed our own. When we put in the second stud mill, which was quite a bit bigger, we had two sides. Of course we had to borrow money from the bank. The bank wouldn't loan money unless we had the stamp of an engineer. They should have stuck it up I don't know where. They made more mistakes in that building when they built it, than anybody could. When they were done most of the work on the mill—they were from Portland—I said "Well, when the mill starts, I'll want the guy that engineered it out here for a week."

That guy had been working for about 15 years I think, and he had never been in a mill he had engineered when it opened. When it opened he says "I can see now why you wanted me out here to show me what I did right and what I did wrong." There was a lot of wrong. He was very happy that I wanted him here. He stayed for a week.

GABE PINETTE

small crawler tractors were used to skid logs relatively short distances to bush mills, which were relocated when skidding distances became too great. With bigger trees this was a cost effective system, but as the trees became smaller, logging costs increased to the point where the Cariboo lumber industry became uncompetitive.

Elsewhere in the world, loggers were encountering similar problems. In some areas where the industry was well established—Scandinavia, eastern Canada, the mid-western states— manufacturers of logging equipment were a part of the business and responded to new needs by designing and producing equipment, some of which was easily adapted for use in the Cariboo forests.

The first machines used in the Cariboo to skid logs over longer distances to mills, which were becoming increasingly stationary, were surplus military vehicles. Four- and six-wheel drive trucks were equipped with a winch and an arch mounted on the rear deck and employed to skid bundles of logs over long distances along snow and ice roads. These vehicles were widely used through the late 1950s and early 1960s.

By the sixties, arch trucks were being replaced with purpose-built machines, rubber-tired skidders such as the Blue Ox and the Timberskidder. Smaller, articulated rubber-tired skidders, such as the Tree Farmer and Timberjack, replaced horses and tractors for skidding logs out of the woods to roadside. From there they were loaded on trucks and hauled long distances—50, 60, even 80 kilometres—to mills that were in fixed locations and, by this time, combined with planers. The new Lignum mill in Williams Lake was in the forefront of this movement.

In the late 1960s, mechanical falling was introduced, first with hydraulic shears mounted on crawler tractors. They were slow and cumbersome machines, however, because they had to drive to individual trees. There was no future for them in the small pine stands of the Cariboo-Chilcotin. In 1968, an entirely new machine appeared on the market, one that was well suited for conditions in the central BC Interior.

A Caterpillar equipped with hydraulic shears was the first mechanized falling equipment used in the Cariboo-Chilcotin.

The Drott feller buncher was equipped with a hydraulic shear and grapple attached to a knuckle boom, mounted on an excavator base. It could cut and pile a 15-metre swath of trees at the rate of about 100 stems an hour. These machines were quickly combined with skidders equipped with grapple attachments that were capable of gripping the bunched logs and hauling them to roadside where they were delimbed and topped, initially by hand with chainsaws, but soon after with a variety of mechanized processors. Then the tree-length logs were loaded and trucked to the mill.

The new technology effectively brought to an end the practice of selective logging in much of the region, as the machines were unsuitable for the type of diameter-limit harvesting used in Douglas fir stands. The advent of clear-cut logging, in turn, required reforestation by planting instead of relying on the natural regeneration methods used previously. Thus, forest management and silviculture arrived in the Cariboo.

An early-model Drott feller buncher on trial with San Jose Logging in the 1960s. The contract logging company began working for Lignum in 1962.

Initially, the Forest Service was responsible for forestry and silviculture work on Crown land, but as the business evolved, forest companies began to take on some of the task, first planning and road layout, then reforestation. Forest companies began to hire foresters to oversee this work.

Lignum's first forester was Andy Szalkai, a former Hungarian forestry student who had fled to BC along with the entire faculty and student body of Sopron University during the 1956 uprising against the country's communist dictatorship. During the summer of 1958 he worked at the Springhouse sawmill and when he graduated the following year, he was hired as a forester.

University-trained foresters were considered useless appendages by loggers and mill workers in those days so Andy's first priority was to gain the respect of the dozen or so mill operators associated with Lignum, demonstrating he knew how to fall a tree or pull lumber off the greenchain. Then he was able to get on with his most important job, finding and bidding on timber. Once that was accomplished, he laid out roads, planned logging and generally facilitated the business of knocking them down and dragging them out. Forestry work consisted of ensuring the stands were harvested in such a way as to provide natural regeneration. In 1965, Szalkai left Lignum to work for Weldwood, where he stayed for another 30 years, until retirement.

In 1969, John Marritt became the company forester. His first job was to chew out a logger for bringing in logs smaller than eight inches top diameter. For many years his primary responsibility was to get the right profile of logs to the mill yard, at the right price, and on time.

By the late 1960s, a new, longer-term form of timber licence, the Timber Sale Harvesting Licence, was available and it required the preparation of a five-year forest development plan. Marritt prepared the first of these. He also was responsible for finding timber and developing the new logging and management systems needed when Lignum eventually moved west of the Fraser into lodgepole pine country.

Centralization of the mills and elimination of bush mills also created a need for full-time loggers capable of all phases of logging, from falling trees to delivery of logs to the mill yard. One of the first such loggers in the Williams Lake area was Dean Getz.

Originally from Saskatchewan, he started logging on Vancouver Island in the 1930s for Gerry Wellburn, founder of the Duncan Forest Museum. In 1953 Dean migrated to the Cariboo, bringing with him a well-used TD6 crawler tractor. He began logging for Paul Trobak's bush mill near Williams Lake. The following year he and Earl Peterson formed San Jose Logging and went to work for Pinette and Therrien. They began by hand falling and skidding with a tractor, then graduated to a new Westfall skidder and a pair of new Mack logging trucks they loaded with a Scoopmobile fork lift. Getz and Peterson began logging for Lignum around Williams Lake and in the Likely area in 1962. Three years later, they divided their assets, with Dean holding onto San Jose Logging and the contract with Lignum. San Jose had its own timber quota for a time, but eventually sold it to Lignum when it became a bone of contention in price negotiations. From that point, the company became Lignum's chief logging contractor.

The transition from logging Douglas fir to lodgepole pine was difficult for loggers as well as the mills. Methods and equipment were unsuitable for harvesting the much smaller pine trees. Falling, delimbing and bucking individual fir trees of 24 inches in diameter, then putting a choker around the logs and skidding them to roadside with a D7 Caterpillar tractor made good economic sense. Utilizing the same system on an 8-inch pine tree was too costly.

In the 1960s forestry was very simple. In drybelt fir, you had a diameter cut and it was designed so natural regeneration was there, so we had very little planting. All we were concerned with in the early days was ensuring that our harvesting was conducted to that cut. We didn't cut any under-diameter trees. Our utilization standards were 50% soundwood. We were allowed an extra inch around the rot. Sweep and crook, there were requirements there. Making sure we didn't rut the skid trails too, the government was always watching us for that. Other than that, at that time, that was it.

We didn't have a problem with the next crop, or size of clearcuts. I remember some large clearcuts in the Springhouse area and talking to the ranger at that time, saying "Jesus, maybe this is too big."

"Nah, don't worry about it!" he said.

Dealing with ranchers was always a challenge for us. We were using the same piece of land as they were, so we had to watch where our cattle guards were, and make sure the fencing's back up. In those days that was a major issue and I remember a number of discussions over whether we needed a cattle guard.

Very complex forestry at that time!

JOHN MARRITT

New equipment had to be purchased. In an era of unwritten contracts and uncertain operating seasons this was risky for contractors. Lignum's solution was to provide or help secure financing for its contractors, with a fixed repayment schedule tied to log deliveries. Negotiating rates in these circumstances was not always easy.

The contractor system worked fairly well, but it was always a delicate situation requiring a great deal of tact and trust on both sides. The relationships that developed in—and survived—this period formed the basis of solid working alliances able to ride out even tougher conditions in the future.

Mill technology underwent the same kind of transformation as had occurred in the woods. In the Cariboo this evolution was partly engendered by the creation of a pulp and paper industry. Prompted by investors, the provincial government concluded that the vast amount of waste created by the sawmill sector constituted a timber supply for a pulp and paper sector in the central Interior.

In the mid-1960s, three pulp mills were built in Prince George, and near the end of the decade one was built at Quesnel by Western Plywood's successor company, Weldwood, in partnership with the Japanese firm, Daishowa. These provided sawmills with a market for their residual materials and logs not suitable for lumber. By agreeing to harvest smaller trees and to utilize more of each tree, down to a four-inch top, they were also able to increase their Allowable Annual Cut. Lignum made this move, installing a chipper in 1963, and was directed to sell its chips to the Canfor pulp mill in Prince George at a non-negotiable price.

A lumber industry organized around many, widely dispersed bush mills and a few centralized planers was unable to take advantage of these new opportunities, which provided an incentive to centralize sawmills. A lot of bush mill operators holding timber rights sold out to the planer companies at this time, electing to take a cash payout instead of going deeply into debt to finance the conversion to higher utilization.

Further incentives were provided by the pulp mills, which were prepared to finance the purchase of debarkers and chippers by the sawmills. The Williams Lake mills, located as they were on the PGE, were in a good position to take advantage of these opportunities as they could ship chips to the pulp mills by rail. This was a factor in the relocation of the Jacobson Brothers mill to Williams Lake in 1964, as their initial location at Horsefly prevented economical shipment of chips to the pulp mills by truck. With rare exceptions, the mills that did not relocate on the rail line disappeared.

The most important technical factor in this development was the invention of the Chip-N-Saw, a radical new approach to processing logs. The idea for the machine originated with Ernie Runyan, an electrician at a sawmill in Shelton, Washington. His company had no interest in his concept so he turned it over to Len Mitten, head engineer at the Canadian Car Company's Vancouver plant where sawmill equipment was built. The first production model was built in 1963 and sold to John Ernst in Quesnel.

The Chip-N-Saw consisted of a linked series of chipping blades and circular saws. The chipping blades reduced a debarked log to a cant, producing high-grade chips for pulp production. The cant continued through the saw where the circular blades cut it into boards. The process produced very little waste, reducing sawdust production from 25 to seven percent of a log. As much as 70 percent of the log ended up as lumber, with most of the remainder as chips.

Perhaps the greatest advantage of the Chip-N-Saw was that it enabled the development of continuous-flow sawmilling. Unlike the industry-standard carriage mill, which moved a log back

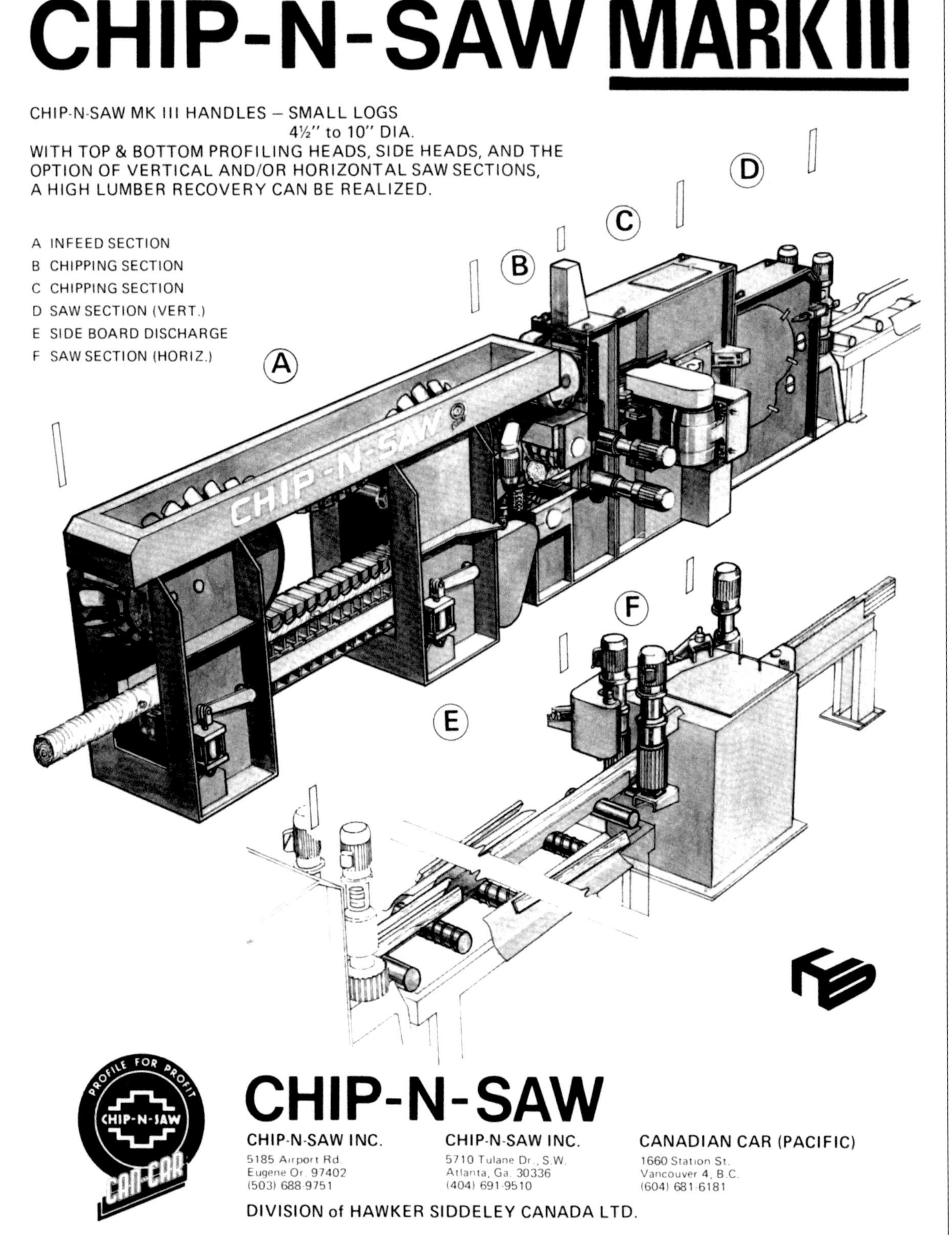

The machine that transformed the Interior BC lumber industry.

and forth through a fixed saw blade, the Chip-N-Saw processed logs in a continuous, forward motion. Conventional gang saws also did this, at a rate of about six metres (20 feet) a minute. The first Chip-N-Saws processed logs at 27.5 m (90 feet) a minute, and within a couple of years were running at 91 m (300 feet) a minute.

Conventional sawmill technology was designed to mill larger logs, from about a 12-inch diameter and up, and to maximize the quality of lumber cut from each log. But these saws took almost as long to cut a small log, which contained very little high-grade lumber. Almost overnight, the Chip-N-Saw provided a means of efficiently milling small logs into standard dimension lumber—mainly two-by-fours and two-by-sixes.

At least, this was the theory. In reality, it took most established mills months, if not years, to fully realize the advantages of the new technology. It was, perhaps, the most radical transformation in lumber making in more than a century and it took a long time for mill workers to adapt to it, and for other components of the mill to be modified to accommodate the Chip-N-Saws.

After Ernst's initial purchase and an installation at the S.M. Simpson mill in Kelowna, the next few Chip-N-Saws built were snapped up by Sam Ketcham at West Fraser. The first one went into his mill in Quesnel, the next one came to West Fraser's Williams Lake mill. The key components for the competitive logging and processing of the Cariboo-Chilcotin's vast lodgepole pine forests were falling into place.

Lignum, in response to these rapid changes, underwent a constant transformation throughout the 1960s. The retreat from Quesnel was followed by growth in the Williams Lake operations. In the beginning, it was not an orderly growth, proceeding from a carefully thought-out strategic plan. Rather, it was a haphazard expansion of operations in response to circumstances as they arose.

At the beginning of the decade, a dozen or more mills owned or controlled by the company and scattered around the region were producing rough lumber for finishing at one of the Lignum planers, primarily the one in Williams Lake. Some of these, like the Kohnke brothers' operation at Tyee Lake, were stable, reliable operators. Others came and went as operators moved in and out of the business, or shifted their allegiances from one planer to another.

As the ranks of these mills thinned during the 1960s, Lignum began to acquire more and more of their assets, primarily timber, but in some cases operating mills. An independent or semi-independent operator might decide he had had enough of the lumber business and offer to sell his mill and timber to Lignum. Or, he might get deep in debt to Lignum and bail out, handing over his mill and timber as payment or partial payment of the debt. These sort of deals tended to occur randomly, not according to any kind of rational plan, and usually when the market was at a low ebb. Lignum was left trying to recover its money from an operation that, in most cases, was not in great shape.

Far left: The Lignum mill in early 1965, with the chipper at left.

Top right: Williams Lake, June 20, 1963, with Lignum burner at centre.

Bottom right: Sawmill at new mill site in November, 1962,

Two such deals left Lignum, by the latter half of the decade, spread thinly over a wide operating area. One was at Tatlayoko Lake, 240 kilometres west of Williams Lake on the western fringe of the Chilcotin plateau, up against the eastern slope of the Coast Mountains. The Tatlayoko mill was acquired from Edgar Hamm in 1961 in a debt-for-assets swap. It cut about 50,000 board feet of rough coastal Douglas fir lumber a shift, which was trucked through the dust and mud to Williams Lake for finishing. Operating costs were high and the timber supply was limited.

The Tatlayoko operation had its own unique character. It was managed by Bob Catcheside, a well-educated, erudite and eccentric Englishman who, even in 1958, wore an earring. He, like much of the crew, had a problem with alcohol and preferred to work in a remote location, away from temptation.

Every second weekend, Catcheside and his crew came to Williams Lake where the lot of them invariably ended up drunk and in jail. On Sunday morning Lignum's general manager of the day would bail them out and dispatch them back to Tatlayoko for another two weeks of work.

A similar acquisition involved a mill at Tappen Valley, located even farther away on the Trans-Canada Highway near Salmon Arm. A sawmill was operated there by Gordon MacInnes. Lignum Sales sold its lumber. When it ran up an unpayable level of debt, Lignum bought it in 1967. A year later, the nearby Salmon Arm Planermills was purchased and when it burned in 1969, the equipment from both facilities was used in a new mill, Tappen Valley Timber. It too had terrible timber supply problems and, in anything but the best of markets, leaked money steadily.

During the 1960s, Lignum's real strength of the company lay in its Vancouver office, in its management and sales division. Lumber sales and marketing were Leslie's strong suit and, from the outset, the sales office earned good profits, most of which were ploughed back into the production division. In 1962, a separate sales company, Lignum Sales, was incorporated. By the mid-1960s, Emil Turner was running it with one other trader, Charlie Marshall, while Leslie spent most of his time dealing with situations in production.

Chip bin at the Tappen mill near Salmon Arm, 1960s.

Lignum in a way is a unique operation. I guess you can call it proof that bigness doesn't mean always that it's good. Lignum managed to remain small and became very efficient because we could see opportunity and were in a position to grab it, to take it. And to even sort of be risky about it. To venture into areas which were really not quite safe for a larger organization, which would think "What is the possibility of success?" and analyze everything to death.

Lignum was in a position to say "Let's try it, let's do it, and see how we can improve upon it as we go along."

Leslie realized very early in life that for a sawmill to be successful then he himself couldn't do it. He didn't have the knowledge, he didn't have the ability, but he had the people. And he was very wise and very clever at picking the right people to do the job. That's really important.

I came in 1966 when the company was in very poor shape. They had a manager who really didn't have it, so to speak. He was good to start with, he was good when the operation was very, very small, but as Lignum grew, he just lost control. So they got in the process of changing the management, and Leslie didn't know how to go about it at that particular time.

It was a case of not having a master plan, but of "Let's try this and see if it works." And if it doesn't work, he had enough wisdom to say "That's enough, that doesn't work so let's give it up and let's try something else."

I was hired as a trader. At that time, there was a salesman, Emil Turner, and another trader, Charlie Marshall, who Leslie thought would be the right guy to manage in Williams Lake. So he shipped him to Williams Lake.

Charlie was a good guy. He was bright. He was very energetic. For him working 12 to 14 hours a day didn't mean very much. That was just fine. So he shipped him to Williams Lake to more or less understudy the existing manager and eventually replace him. Keep in mind that the manager at the time had 10 percent of Lignum, he was part owner.

I came aboard to help Emil to do the wholesale trading and sell the production of Lignum. That didn't last very long. I don't think I was in the trading office more than maybe two or three weeks when Leslie approached me to go to Williams Lake to see if I could give Charlie a hand.

It was really a mess, with an accountant who was half of the time drunk. They had absolutely no office routine. Contractors lining up on the 15th and the end of the month to get paid, and we were busy getting information together about how many truckloads they brought in. It was just an unbelievable situation.

I asked the manager, Gordon Bruce, if there was any kind of record what these contractors are supposed to be paid. "Oh, no, just play it by ear." So it took some time to get the thing really going.

I thought the first thing was to put some order into this setup so that they would know what they're supposed to be doing, and put some kind of costing control in so it runs like a company should. I had to clean house. I couldn't work with these people, they were so accustomed to being so sloppy it just didn't work.

Eventually I was instrumental in persuading Gordon Bruce to sell off his 10 percent and leave, and Charlie became the manager. I stayed with him another 18 months, and then I came back to Vancouver.

☙

SIDNEY EGER

At about this time, Marshall was dispatched to Williams Lake to work under Gordon Bruce and subsequently to replace him as general manager. A new trader, Sidney Eger, was hired to replace Marshall.

Sidney had come from Czechoslovakia a few years earlier to work on a turkey farm, but when he arrived in Vancouver the farm had gone broke. Instead, he went into the forest industry and worked in a number of locations around the province before joining Lignum.

After two to three weeks in the trading office, Sidney was sent to Williams Lake to help Charlie reorganize the production operations. He was there a year and a half before getting back to Vancouver. By this time he had earned his nickname, the Eagle, for his sharp financial eye.

At this stage in its evolution, Lignum's production division had reached a hiatus. The company's Cariboo operations had begun as a simple planing mill. The mechanics of this type of operation were not complicated, and managing it was not particularly difficult. Leslie, although he understood the operation of far more complex mills than the one at Williams Lake—he had managed Alaska Pine's large coastal mill for several years—was not mechanically inclined. When it came to the details of mill construction, operation and modification, he had to defer to others. This was where he encountered an obstacle that frustrated his plans for the rest of his life.

The lumber industry was relatively new in the central Interior, and had not developed a large number of skilled mill operators and managers. At this stage in the industry's development, technological change was speeding up, along a different route than that followed on the Coast. It was not difficult to find a production manager who knew how to run an established mill of that era, particularly if he had built it. Leslie was great at working with these people, setting up the financing, selling the lumber, and so on. But when it came time to expand or upgrade the operation, and the manager was not up to that task, Leslie was stymied. He did not understand the mechanics of mills himself to the degree required to make the necessary improvements and, in spite of many attempts, was unable to find a forward-thinking manager capable of taking his vision and running with it. During this period the production division went through a number of general managers, all of them good in their own way, but none of them equal to the task of restructuring the division into an efficient, coherent and—compared to the sales division—profitable entity.

Leslie Kerr in 1963.

Gordon Bruce, who also owned a share of the company, was perfect for the task of developing the planing mill and establishing relationships with bush mill operators, employees and others essential to the success of the business. But when improvements were needed to keep the mill competitive, and more rigorous financial management required to keep it profitable, other skills were required.

This was the task assigned to Charlie Marshall, and once he was established at Williams Lake, Gordon retired. He sold the interest he had earned in the company back to Leslie. Charlie was a good financial manager and administrator, but at heart he was a salesman, not a mill man, and after two years he left to manage Holding Lumber at Adams Lake. At that point, Leslie began looking outside the company, and outside the region.

The next manager hired was Bob Waters, who came from Canadian Cellulose in Prince Rupert. This was a big, well-established corporate operation, and did not provide the entrepreneurial experience needed to run an operation like Lignum—small, but with the potential for growth. Waters lasted six months before being replaced by Herb Dell, one of Lignum's first employees at Quesnel.

At this time, Gene Rolston was hired to oversee installation of the new Chip-N-Saw. Rolston had learned the sawmill business in Europe and come to BC as part of the Koerner exodus, working for many years in coastal mills, most recently at Alaska Pine and Weldwood. He was a rough, tough old sawmiller whose management style leaned toward the kick-in-the-pants school. He knew mills up, down and sideways and, although he was not familiar with Interior mills, understood their needs. He had been in retirement for almost five years when he came to Lignum. Perhaps his major contribution was to bring in Mike Madrigga to help oversee conversion of the mill operation from a conventional fir-spruce producer to a small-log mill utilizing a higher proportion of pine.

This done, Gene went back into retirement, and Madrigga replaced Herb Dell as manager. He left to work for Northwood after a couple of years, and Charlie Marshall came back for a year or so, until Reid Banford, a long-time Interior mill manager with a fearsome reputation, was hired for the job. These constant changes in senior management, during a period of rapid technological adjustment, did not make for a particularly efficient and profitable production division.

The glue holding the production operations together was a handful of long-term employees, most of them in Williams Lake, several of whom had been with the company from the outset. Stew Smith was plant superintendent and assistant manager during much of the 1950s. Pete Routley spent 31 years as the mill's carpenter foreman. Cliff MacIntosh worked as planerman and planer foreman from 1948 to 1963. Wendell McDonald managed the Likely sawmill until it was sold, then served as mechanic shop foreman in Williams Lake until 1969. Art Goyette ran several bush mills for Lignum and was assistant manager of the Williams Lake mill from 1961 to 1966. Roy Astley started out building the mill, worked as lumber grader and first aid man, and ended up taking care of the dry kilns before retiring into the job of company historian. These men, and a few others besides, laid the foundation for Lignum, held the mill together through a chaotic decade or so and trained their successors. Without them, it is unlikely the company would have survived the lack of clear production management in the 1960s.

In the meantime, Leslie struggled to make a coherent operation out of a mismatched collection of mills spread across the central Interior of the province, without the financial means and the manager he needed to build the modern operation he knew was required. As a result, the company's production division lacked a sharp operational focus and the ability to invest strategically.

Good workers were hard to find in the 1960s. New technology required increased skills and the highly transient workforce employed in the lumber industry at that time did not possess those skills in abundance. Situated as they were, midway between Vancouver and the larger mill centre of Prince George, the Williams Lake mills suffered from a high turnover of workers, some of

---

At the end of the day, how to log really came from the loggers. They were the ones that developed it, with the equipment people. Trying prototypes and bringing in new equipment to see if this was better than the old one. When the equipment is changing you have to develop the mechanics and new operators to deal with the new type of equipment, so it was not only a mechanical thing, it was also a people thing that evolved, and the changing attitude towards it.

It just seemed to work out. Moving with your contractors to a new piece of equipment seemed to be the right thing to do. Sure, there was a lot of consternation some days when we tried to move a fir logger into pine. We learned we've either got to set him up for pine as well as fir, or move him over to pine 100 percent. It does take different logging equipment. There was an evolution there, of changing ideas and methods. But Lignum always worked very closely with the logging contractors, and it just evolved. It just seemed like a natural way to do business.

We used to have a lot of disagreements over logging rates and the trucking rates. It was never enough. The truckers are another thing, mostly independent truckers. San Jose had their own trucks, but we also had a long list of independent truckers. Don McDonald was our number one trucker.

I remember a number of days when our mill was just about out of wood. It was breakup, the roads were muddy, we desperately needed logs and boy, the loggers and truckers just bent over backwards to be sure we didn't run out of logs. I truly admired their cooperation, the effort that they put into making sure that we had logs at the right time. They were pushing and pulling trucks that they'd just spent a whole bunch of money on. They worked long hours, the whole weekend, twisting wrenches, fixing their trucks so they're ready to go Monday morning.

JOHN MARRITT

Top: The July 8, 1963, fire when sparks from the sawmill burner destroyed 400,000 board feet of lumber at Lignum's Williams Lake mill.

Bottom: Lignum's mill yard sign after the move to the new Williams Lake site in 1961.

whom would stop off and work only long enough to buy gas to complete their journey to one or other of the major centres. One month, Sidney Eger was astounded to discover that, while the mill required 150 workers to operate, cheques were issued to 600 people.

Williams Lake, when I left, was at 200,000 to 250,000 per day production. They had a gang saw, a sash gang. It shook itself loose and blew up every so often. I had more trouble with the insurance company, Jesus Christ!

The mill had a head rig, sash gang, they had a scrag mill and they had a cant edger behind that. They had a board edger and all the lumber went on to a greenchain. They had umpteen million people pulling on it. At that time, it was State of The Art! God, I remember the payroll, you wouldn't believe it. We must have had at least 600 cheques, payroll cheques each pay period. People were coming and going, going and coming. You never know who's going to show up!

Part of it was due to the fact that it was at Williams Lake halfway between Vancouver and Prince George, so people would stop for a few days, just to earn a little money and they went on. Later on, that greenchain was ripped out and we put in an edge sorter. I don't know if you remember what edge sorters are all about. The lumber was stood on edge on long, long belts running along. As the lumber went along, the two-by-fours kicked out, the two-by-sixes were kicked out and so forth. When they were kicked out, they were piled by people pulling lumber. It was a Workmen's Compensation nightmare. You can imagine these pieces going so quick and dropping off. Sometimes they didn't drop fast enough and another piece hit them, so it wasn't the safest operation. But that was the State of The Art at that time.

Sidney Eger

Meanwhile, in Vancouver, the sales division was also experiencing changes. When Sidney went to Williams Lake to impose some much-needed financial discipline and another trader was needed, Ken Ross was hired. He had obtained a university degree and a teaching certificate before discovering about the existence of the lumber trading business, which looked to him like a far more interesting occupation. Once he'd tried it, he gave up any further thoughts of teaching and went to work for Cooper-Widman, perhaps the best known wholesale lumber trading firm in the province at that time. Its operations and organization served as a model for others created in its wake, most of them staffed with Cooper-Widman alumni.

Ross brought with him a good knowledge of the North American lumber markets and eventually he brought in other Cooper-Widman-trained traders familiar with that market. He had opened and run Cooper-Widman's New York sales office and had a good understanding of the complexities of the US market. By this time, Lignum had more or less abandoned the European market and was selling heavily into the US.

The evolving US market was primarily for dimensional lumber, an undifferentiated commodity. Development of the North American rail systems during the 1950s and 1960s, coinciding with the growth of the Interior BC lumber industry, made it possible to ship carload lots of lumber, more or less directly from the mill to customers throughout the US. Before this time, BC lumber shipments to the Atlantic seaboard were primarily from coastal mills on ships, and consisted of much larger orders to wholesalers in eastern US ports.

The advent of mills such as Lignum's at Williams Lake, and the growth of rail transportation, created a new type of lumber market with new procedures and protocols. By the early 1960s, there were scores of mills throughout the Interior selling more or less standard-dimension lumber to hundreds of buyers throughout the US, except for the Pacific Northwest. Few mills had their own sales departments, electing instead to sell through wholesale traders such as Cooper-Widman, or companies such as Lignum, which sold a mix of their own and others' lumber.

The fact that practically all this lumber was milled to standard dimensions instead of to a customer's specifications created a classic free market of many willing buyers and sellers. Demand was created, primarily, by the construction of new houses, a fickle market at the best of times, based on largely unpredictable decisions made by hundreds of thousands of individuals and families each year.

Lignum began wrapping lumber about 1967.

The mills, with no direct knowledge of current market conditions, focussed on producing as much lumber as possible. The lumber trader's or sales agent's job was to move it out of the mill yard as quickly as possible. Mill managers did not like to see growing inventories of lumber in their yard and were willing to negotiate prices with buyers from the wholesale trading houses.

Lumber traders, especially in pure wholesale houses, began as field buyers, spending most of their time travelling from mill to mill chatting up the managers and purchasing lumber. There was fierce competition among wholesalers to capture mill business and loyalty, making personal relationships between buyers and mills' sales managers important. A trading operation such as Lignum Sales relied less on buyers, obtaining most of its supply from the company's own mill, or from other mills with which it had ongoing sales agreements.

Eventually, field buyers would move into the office to learn the sales end of the business. Orders from established customers were received and compiled in a list, and orders for lumber already purchased or known to be available were vigorously solicited by the sales staff, mostly by telephone or telegram. Some orders were sold "back-to-back," matching a mill's production with a customer's requirements. Others were accepted with no lumber to fill them. Some orders might need filling at more than one mill, or require special milling or treatment procedures. A wide array of permutations and combinations of purchases and sales was possible, necessitating complex negotiations and transportation arrangements.

The phenomenon of "transit sales," unknown outside North America, evolved. A trader dispatched an unsold boxcar of lumber from the mill to one of a number of transfer centres in the US, via the most indirect route or slowest carrier available. The car would slowly make its way to the transfer centre while the trader attempted to sell it. It might travel around the country for a month before it was sold, at little or no cost to the shipper. If the shipment was not sold, the car would end up in one of several storage yards scattered throughout the US, where it would become subject to demurrage, a charge for failing to unload the shipment within a specified time.

Top: The mill yard from the south, November, 1962, with the sawmill at centre and log deck at left.

Bottom: A 1960s argument between a forklift and a welding truck.

At any given time, a trading office such as Lignum might have 100 or more cars of lumber in transit. Traders knew exactly where each car was at a given time, and worked hard to sell them before they reached their transfer centre. If they failed, the lumber would have to be sold quickly, probably to a sharp-eyed US trader who understood the situation. Eventually, as trucking of lumber became economical and the railways realized one of the primary functions of their boxcars was to provide free inventory storage for the traders, they put an end to the practice.

In the wholesale lumber trade, there is no fixed price for lumber at any given time. Price is whatever the traders agree on in any specific transaction. Two minutes later, in another transaction, it may be a different price. If a trader thinks the price will increase tomorrow, he might stop selling today and go play golf. Tomorrow he may sell lumber he has already bought at a higher price than he can get today.

Lignum differed from most trading houses in that it sold its own lumber as well as that of others. Until the early 1980s, about 75 percent of lumber sold came from Lignum mills, and 25 percent from outside. This kind of setup creates a potential conflict of interest for traders, who normally work at one rate for selling in-house wood and another for the rest. If a trader knows the price will rise tomorrow, whose lumber does he sell today? The one that earns him the biggest commission, or the one that produces the largest return to the mill?

Traders and mill managers often fail to see eye to eye. In many cases, the number one priority at a mill is volume. There are all sorts of prizes and rewards for the crews that set new production records. Cutting to meet a market strategy based on the needs of particular customers was unknown until a decade or so ago. This was more so in the past than at present, but some companies still operate on this basis.

Wholesale traders are not excited by the volume of lumber a mill churns out; traders are excited about the net profits their sales earn. Traders have all sorts of ideas about how mill managers can increase net profits by changing the profiles of the lumber they cut. Mill managers, whose priority is production, are rarely enthused about these ideas. They are more concerned about moving finished lumber out of the yard, at the lowest cost and highest speed possible. Good lumber traders and good mill managers are often quite different kinds of people and view each other with suspicion, if not disdain. The task of senior managers in this sort of situation is to build a lumber trading division that works in close tandem with the production division, rather than against it. This undertaking began at Lignum in the 1960s, when the foundations were laid for the tightly knit organization that later emerged.

It was during the 1960s that Leslie and Barbara's sons began their association with the business. The couple had three children: John, born in 1944; Tim, born in 1946; and Pat, born in 1953. From an early age, the boys accompanied Leslie on trips to Williams Lake. The ever-changing fortunes of the company were the subject of dinner table discussions at the Kerr home near the University of British Columbia in Vancouver's West Point Grey district.

In those days, it was expected that sons would become part of the business as they grew up, and that daughters would not. John and Tim worked in various jobs during summer breaks while they were going to school. Pat, regardless of her views on the matter, did not. As teenagers the boys spent summers working at the Williams Lake mill and in the woods. John lived with the Bruce family for two summers, and with Felix and Maisie Kohnke for another two. They knew from the outset this was a tough, uncertain business.

During this period of uncertainty, when the future of the company was occasionally in jeopardy, it was unclear to Leslie, as well as the rest of the family, whether he was trying to build a family firm with a long-term future, or if he was putting together a property that could be sold for cash at a propitious moment.

Selling a lumber business at a profit is probably more difficult than building a profitable company. A forced sale, when everything has gone wrong and the bank has a gun to the owner's head, takes no particular skill. To sell and walk away with a bundle of cash is an opportunity that comes along only occasionally, and is usually not obvious when it does. Fortunately for the survival of the industry, lumbermen are always optimistic about the future value of their businesses.

Given the character of the 1960s, John and Tim began to develop their own ideas about what they might do. After John graduated from the University of British Columbia, where he acquired the nickname "Jake"—by which he would become known in the industry—he decided to do graduate work in the US. Leslie persuaded him to avoid the Ivy League colleges he was eyeing and enroll in marketing and finance at the University of California in Berkeley—just as it was on the verge of turning into a hotbed of political ferment and student radicalism. Leslie figured Jake would learn more there than at Yale or Harvard.

Jake came back to Vancouver for a summer at Leslie's request to take a close look at the state of the company. After a few months, he conveyed to Leslie his opinion that it should be sold. Then he headed back to the excitement of Berkeley.

Jake thrived in that atmosphere, and after obtaining a master's degree, went to work as a market planning manager for International Paper in New York. From there, he took a job as account executive in Botsford Advertising in San Francisco. In the late 1960s this was, perhaps, the most exciting city for a young man of Jake's sensibilities to end up. A massive cultural revolution was underway and San Francisco was its epicentre.

Tim stayed in Vancouver, immersing himself fully in the opportunities and temptations offered his generation during the 1960s. He worked at various jobs within Lignum, starting out with Ken Ross trading lumber, and worked at several tasks in the production division. But Tim was too independent and intellectually adventurous to work comfortably in a family firm, and by the end of the decade he had gone off on his own to learn the real estate business.

A few years earlier, in the mid-1960s, Leslie had experienced a heart attack. He survived it, and by the end of the decade was thinking seriously about retirement. His friend, Laci Kardos, had long since sold his company and was enjoying retirement, as were many others among his contemporaries. Leslie was an active member of the Shaughnessy Golf Club and found he was enjoying himself more there than in travelling to Williams Lake to solve the latest problem with the sawmill, or whatever else demanded his attention.

He asked Jake to come back and help him put Lignum into saleable condition. Reluctantly, Jake agreed.

CHAPTER 5

THE NEW GENERATION

As Jake was about to learn, running a company like Lignum is comparable to juggling live cats. There are essentially three components to the company, each staffed by a semi-anarchistic collection of independent-minded people. The more capable they are, the more difficult they are to manage. Lignum has always sought out the best people in their fields, and quite often found them.

When Jake came into the company, the woods end consisted primarily of contract loggers. Logging, from a distance, looks like a simple task. You cut down trees and deliver the logs you make from them to the sawmill. In reality, it is an incredibly complex task, beginning with the fact that every tree in the forest is unique. Logging is not a simple, repetitive process. It requires ingenuity, intelligence, stubbornness and—on occasion—guile. No logger ever believes he is paid what he deserves, or if he does he never lets on. Most of the time, good loggers are able to conceal their basic contempt for mill people, lumber salesmen and other lesser beings—including foresters and corporate CEOs—who have any influence over their lives.

At the other end of the operation is the sales division. Wholesale lumber traders occupy an arcane world of their own in which the primary task is to play all ends against the middle. They must maintain a balance between generating sufficient revenue for the company to prevent their shop from being closed and the task of selling lumber farmed out, and earning more in commissions than their soulmates who become rich hustling mining stocks on Howe Street.

A good sales division, such as Lignum has had for much of its existence, complements a production division, multiplying its earnings and cushioning its losses. It provides the intelligence needed to survive and thrive in a marketplace that functions in a state of constant flux. A sales division that gets out of hand, as has also happened at Lignum on occasion, can sink a company overnight.

Like Leslie, classic lumber traders have a touch of the gambler in their blood. Their working— and in many cases their private—lives consist of an unending series of calculated risks. The best sales people are as apt to be as contemptuous as loggers about co-workers in other divisions of the enterprise—especially of senior managers and owners who are not willing to bet the farm on a daily basis.

On the surface, mill people are the sanest of all. They are steady workers, not given to the sort of manic-depressive behaviour exhibited by loggers and lumber salesmen. Day after day they patiently cut logs into lumber, load it onto rail cars or trucks and watch it disappear. They are well organized and work according to clearly defined rules laid out in signed agreements between their company and their union.

Williams Lake Tribune coverage of the dramatic fire in Lignum's log inventory in 1969.

Senior mill people often have enormous egos, which may derive from the fact that they produce the product that generates the company's revenues. They are enormously proud of their role in the company, and at their happiest when setting new production records.

The difficult thing about mill people is that the better they are, the more money they need for new and improved equipment. Good mill people are always thinking of ways to do their job faster, easier and more profitably. They are always confounding owners and accountants with technical terms—edgers, optimizers, trimmers, head rigs, scanners. Their favourite place to discuss new ideas is in the thick of the action, yelling to make themselves heard.

In companies like Lignum, there is a tendency among those who work in the different divisions to consider their roles as primary, their contributions to the whole effort as paramount. Loggers usually believe they perform the hardest, most difficult task within the company— harvesting timber under often-treacherous conditions and getting logs to the mill at an impossibly low cost. Once logs have reached the mill yard, they feel, the business of converting them into money is simple.

Mill workers like to think their work of converting logs, the quality of which is beyond their control, into saleable products constitutes the essence of the business. It is their skill and ability to build and maintain a mill to produce lumber which, with very little additional effort, practically sells itself.

Those employed in the sales division, especially wholesale traders, lean toward the idea that the company could survive just fine without its logging and sawmill divisions. Lumber, especially dimension lumber, could be easily purchased from dozens of mills, usually at lower prices than the company mill charges. Situated as they are, closest to the money, they sometimes think the best way to grow the company would be to sell off the mills, increase the sales staff and raise commissions.

In their more thoughtful moments, all these people realize that the success of the company lies in maintaining a fine balance between its various components. This is what general managers and owners know, and their primary task is to maintain that balance, adding strength and resources to one division when it lags, and restraining others when they get too far ahead of their colleagues.

Leslie never had to deal much with loggers. During the early part of his tenure at the helm, the bush mills provided their own logs and delivered rough lumber to Lignum. These people he understood, and he worked well with them. He also knew that the important thing in the woods section was to maintain a secure timber supply. It was his foresight in this regard that provided Lignum with a solid foundation.

Sales was Leslie's great strength. He never had to rely on a room full of traders engaged in an activity he did not fully understand to keep that end of the business healthy. He grew up in the lumber trading business and understood it intuitively.

The mills caused Leslie more frustration than any aspect of the business. He understood them, having managed a large coastal sawmill, but he was not mechanically inclined. Unlike some of his contemporaries, such as Harold Jacobson or Gabe Pinette, he could not roll up his sleeves and rebuild his mills to meet the needs and opportunities he saw. He had to rely on others to actualize his visions and the kind of people he needed, people who could design and build a mill to efficiently process the uniquely different timber profiles of the Cariboo region, were in short supply.

In Leslie's experience, mills were financial black holes, requiring capital that could better be spent on timber acquisition. Mills like the one at Tappen consisted of a ramshackle building full of machinery that broke down regularly. If logs and enough money were stuffed in

one end, lumber came out the other end, only occasionally at a profit. He knew he needed modern facilities, but he lacked both the money to acquire them and the right person to put them together. For most of his tenure, his approach was to spend as little as possible on the mills, while putting every dollar he could into buying timber. In retrospect, it was a wise choice, but it did not always satisfy the mill people.

Jake came into Lignum as vice-president with responsibility for investments and new ventures. In his mind, the primary task was to get the company in fit condition for sale. Although his homecoming could not exactly be described as the return of the prodigal son, he did not see himself as a member of the second generation of Lignum owners taking the helm and steering it into the future. Liquidation of the company to provide for his parents' retirement and a return to the good life in San Francisco within six months—a year at the outside—was his objective.

The whole matter of inheritance is a dodgy one in the BC forest industry. Since the industry began in the 1860s, many prosperous companies have withered when the children, usually the son, of the founder took over. It has always been a worry for the individuals who own or control the industry's family-run firms.

When H.R. MacMillan and his contemporaries were confronted with this situation 25 years earlier, they found it one of the most difficult they had to cope with during their long careers. There was no shortage of examples of natural heirs proving inadequate for the task. In MacMillan's case, having no sons to succeed him, he chose an heir from his company's ranks, treated him like a son and, at the final fatal moment, fired him in a pique of anger. From that point on the company slid downhill to its final demise 40 years later.

> *We ran Tappen with no quota. The best thing we did at Tappen was when Dad gave Sidney $100,000 to go out and buy land with standing timber. We cut all the standing timber on it and in 1990, Sidney and I sold the land for $500,000 or something. Tappen was the orphan where you didn't have quite enough money. We'd make money out of Tappen maybe two months out of ten.*
>
> *You'd get a really good lumber market and Tappen would make money for a year, a year and a half, then you'd realize that you'd stopped making money 'way earlier than in Williams Lake and start making it 'way later.*
>
> *I guess it was at that point in time that we made a conscious decision that we wouldn't expand and buy other places. That's when we sold Tappen and said we were going to focus on one place—Williams Lake.*
>
> TIM KERR

The situation confronting Leslie Kerr and his family was one which, sooner or later, had to be faced by all the family-owned firms in the Cariboo. Few, if any, of them had reached a size by the 1960s where they could underwrite their founders' retirements and, at the same time, permit the founding families to retain control of the company. At about this same time, for example, the Pinette and Therrien families were dealing with this situation.

The three members in that company's founding generation were all approaching retirement. Gabe Pinette's son, Conrad, had come into the business and was gradually assuming management of it. One of his chief preoccupations was to provide a comfortable retirement for the family members who had built the company. He needed money to buy their shares, as well as to expand the company and ensure its survival. His choices at the time were to go into debt or find a financial partner. He ended up selling half the company to BC Forest Products, which contributed timber holdings to expand operations and cash to buy out the founders. Conrad continued running the company, as well as BCFP's other Interior and Alberta operations.

At a later date, when the Jacobson brothers reached this point, and none of their heirs or other shareholders wanted to carry on the business, they sold outright to a family operation from the Okanagan, Riverside Forest Products. North of Quesnel, the six-member Dunkley family, founders of a mill at Hixon, found themselves in similar straits. They were not willing to sell to one of the big Prince George companies that only wanted the timber and would have closed the mill. They looked around and found another six-member family, the Novaks—who had recently slipped under the Iron Curtain from Slovenia—and struck a deal to gradually give them control of the operation. At West Fraser in 1977, when Sam Ketcham died suddenly in a helicopter crash, he was replaced by one of his brothers until his nephew Hank, who joined the business in 1973, was able to take over.

In September, 1968, Leslie sent Jake a financial overview indicating Lignum had excess funds of $600,000 on hand. "The above analysis indicates," he concluded, "that we can safely invest $500,000 minimum in a sound and promising venture, or ventures, at the earliest opportunities."

One of the first things Jake did after he arrived at Lignum was to show the company books to his father-in-law, a senior banker in San Francisco. Jake was shocked to find out that, in spite

Top: Lignum forester laying out cutblocks near Likely, 1973.

Bottom: Skidding to the landing with a Caterpillar worked well with larger fir and spruce logs, but was unsuitable for small pine logs.

of these cash reserves and having shown a profit almost continuously since its formation, Lignum was virtually unsaleable.

The company's production division was thinly spread from Tatlayoko Lake to Salmon Arm, with long distances separating the well-worn mills in these two centres. In 1969, it employed more than 400 people and produced about 100 million board feet of dimensional lumber, 90 percent of which was shipped by rail to the US and the remainder sold in Canada. Sales that year were $6.1 million.

Two-thirds of this production and half the employees were in Williams Lake, while production costs at the other two mills were significantly higher. The operations outside Williams Lake were dragging down the whole company.

A couple of years earlier, Jake had advised Leslie to get rid of the Salmon Arm and Tappen operations. Leslie, however, was reluctant to cut his losses until he had recouped his original investment or proven to himself this was impossible by going deeper into debt.

Management of portions of the production division was still a weak link; there were people capable of running the various operations, but no one with the ability or the mandate to rationalize the entire division had appeared. During Jake's early period in the company, Gene Rolston and Mike Madrigga were engaged installing the new system in the Williams Lake mill to cut small lodgepole pine logs.

The improvements consisted primarily of acquiring a ten-inch Chip-N-Saw for the then-astronomical sum of $245,000. Lignum had never before invested so much in one shot. It was installed early in 1971, and that year profits promptly disappeared. Production costs at the Williams Lake mill topped those at Tatlayoko by $4.37 a thousand board feet.

There were two kinds of problems associated with the installation. First, the crew, most of whom had learned the sawmill business in the old mill, had to operate the conceptually different Chip-N-Saw. Second, the rest of the mill had to be reconfigured and adjusted to accommodate the new equipment. This was a slow and costly process.

By the time Jake arrived, Sidney Eger had risen to the position of controller with primary responsibility for financial planning and overall supervision of the Salmon Arm division. Gradually, he was getting a handle on the company's financial management. Ken Ross was in charge of Lignum Sales, and had brought in Dave Verchere and Wally Pierce, two other Cooper-Widman alumni, to supervise production and sales co-ordination.

Faced with the difficult job of selling the company, Jake began laying the groundwork of a two-pronged strategy. He widened the search for buyers, and began planning a rebuild of the company's production division to make it more saleable.

Leslie had attempted to sell Lignum before this time, on one occasion to Merrill Wagner. Negotiations were completed and the deal was all but closed. But it did not guarantee the jobs of employees, and when Leslie went in to sign the final papers he could not do it, so he backed out.

> *What sets Lignum apart from a lot of companies that started at that time was that Lignum hasn't expanded into a multi-divisional operation, but stayed with one operation. I remember having a discussion with Jake, after looking at various operations, coming to the conclusion "No, that's not for us." After doing this many, many times we came to the conclusion that being bigger isn't necessarily best. Being the best is much better, the philosophy that we should take our talent and our resources, and focus on one operation rather than diverting our talent and our resources into a multi-divisional operation.*
>
> *We looked at a number of operations, there were a lot of opportunities, and we were never comfortable with proceeding with them, for one reason or another. It was, at that time, a lack of capital or it wasn't the business for us. The experience we had when we did expand into the Tappen operation was a drain on management talent and other resources. When you do that, it takes you away from your bread and butter operation, which was always at Williams Lake.*
>
> — JOHN MARRITT

Now, however, he was prepared to see Jake complete the task.

Improving the condition of the production division entailed a number of approaches. The perennial problem of timber supply was as acute as always in 1969. The long-term supply for the Tatlayoko mill was uncertain. One option briefly considered was to maintain Lignum's quota position in the Chilko sustained yield unit but relocate the mill farther north, perhaps in Vanderhoof, where timber was available. This was likely one of the options Leslie had in mind for the surplus cash on hand.

By the time Jake's allotted six months has passed, he was deeply immersed in rebuilding the company and had put off his return to San Francisco indefinitely. Life in the BC lumber business, it turned out, was as interesting as the California advertising game. In the late summer of 1972, his task became a lot harder.

On August 30, the Social Credit government, which had held power continuously for 20 years, went down to defeat at the hands of the New Democratic Party under Dave Barrett. The "socialist hordes" had seized the levers of power, including life or death authority over a forest industry that was almost totally dependent on Crown timber to feed its mills.

Premier Dave Barrett (Left), with John Ailport, during a 1975 visit to Lignum's mill at Williams Lake.

During the campaign Barrett and other NDP candidates had made vague references to expropriating forest company assets, and at their election night celebrations, forest-minister-to-be Bob Williams vowed, "We've been given a mandate and we're going to clean house." No one, including the NDP, knew what lay in store for the forest industry but to most people involved in the industry, the future did not look promising.

The following Saturday, the Russians, whose totalitarian government and regimented way of life Leslie had left Europe to escape, triumphed over Canada's national hockey team. On Sunday, Leslie died at his home in Vancouver of another heart attack. He was 66.

Ownership of the company passed to Barbara and their three children. Jake, at age 27, was appointed managing director and took over leadership of the company. Pat, with no involvement in the company, later sold her shares to her brothers. Tim, aware of the burden of responsibility thrust upon Jake, returned to the fold and offered his help and support.

The probability of two young men in their twenties taking over an ailing forest company and surviving—let alone prospering—in the rough and tumble world of the Interior BC forest industry was not worth betting the bank on. Although they had both grown up on the fringes of the industry, neither had been groomed for the task they now faced.

Both of them stood out in the industry like sore thumbs. The range of political opinions and personal styles within the industry at this time was very narrow, the influence of the cultural revolution not yet having penetrated the BC hinterlands.

Between them, Jake and Tim pretty well covered the range of counter-cultural opportunities offered in the late 1960s. Jake was a sophisticated urbanite, fresh from the hip, cool worlds of New York and San Francisco. He may have worn a suit, but in any boardroom or bank manager's office, his tie was always the widest and his hair the longest. His opinions were definitely his own, and he had no trouble expressing them.

Tim, who had not quite dropped out of the mainstream, had decidedly turned on and tuned into what was happening around him. In appearance, with his long hair, beard and clothes purchased in boutiques along Vancouver's Fourth Avenue, he fulfilled every fearful fantasy of the latest threat to western civilization—a "hippie." News that the Kerr boys were now in charge of Lignum spread through the company, and the industry, like a crowning forest fire. As photographs from industrial archives of that era attest, they were a unique phenomena.

In the late 1960s, Jake was away. I finished university, and I had gone to work trading lumber for Dad. I was the junior guy for about a year and a half, gone on a couple of buying trips to Prince George, looked for mills with beehive burners; you could go along the highway and see the smoke. That's how you found out where new mills might be.
I did a bit of that work for about a year and a half and I was just not of the right mind to work for my dad at the time. It was the end of the '60s. I don't care who anyone was, working for your dad was really hard in those days so I stopped.
I went off and was in the middle of getting my real estate licence when my dad died. I remember thinking that it wasn't fair that Jake was left with the burden of all this stuff. I knew I could be useful, so I told him I would like to come back and work with him. He had no trouble with that.
I went up and worked at Tappen Valley Timber in Salmon Arm, and after about two years there I went up and worked at Williams Lake, in the forestry end.

TIM KERR

The Cariboo Lumber Manufacturers' Association directors in 1976.
Rear (Left to Right) Joe Miyazawa, Fred Linde, Joe Komori, Glen McMillan, Dave Ainsworth, Dave Balison, Ike Barber, Marv Kempston.
Front (Left to Right) Sam Ketcham, Harold Jacobson, Jake Kerr, Conrad Pinette, Jack Newman, Bob Smith.

Their settling-in period was not entirely smooth. At one point, over the mill manager's reluctance and Jake's scepticism, Tim insisted a couple of his university-era friends be given jobs at the Tappen mill. Having no previous mill experience, they were hired to work on the green-chain at Tappen, on the night shift.

This was in mid-winter and, like all Interior mills in those days, the Tappen greenchain was open to freezing temperatures and cold winds. The morning after their first night on the job, the manager phoned Jake to report the new mill workers had led a wildcat strike over what they considered to be intolerable working conditions.

As the Kerr brothers soon learned, however, they were not on their own, but had a loyal and resourceful group of employees to back them up. After Leslie died, the mill people put together a half-million dollar list of improvements they felt were needed. The list included improvements on the beehive burner required by new pollution-control regulations, without which the mill could not operate.

Jake had obtained a verbal commitment from Lignum's long-standing bank, the Toronto Dominion, for the money. The bank's local manager said he needed approval from Toronto. Expecting no problems, and needing the burner improvements to continue operations, Lignum paid for the improvements out of operating capital.

When the Toronto Dominion head office in Toronto learned of this, it refused the loan. Lignum faced a financial crisis and Jake scoured the city for money. Eventually, he obtained a loan from a US bank, and then moved the Lignum account to the new Bank of BC—where, a few years later, he would be appointed a director. He had weathered his first financial emergency.

In the end the only person dissatisfied with the deal was Lignum's sales manager, Ken Ross. He had taken the local Toronto Dominion manager out on his boat for a day of salmon fishing before the rupture occurred. The guy caught a lot of fish and kept them all for himself. Ross had hoped, in vain, to obtain the return of some of these fish.

During this same period, Jake was fielding enquiries and considering offers to buy the company. The only serious offer came from West Fraser, but it was so low as to be almost insulting. After much discussion with the Ketcham clan during a late-night session in a Vancouver hotel, Jake raised the question of his company car. It was his first company car, and he liked it. The Ketchams insisted the car had to come with the company. Curious about how tough they would be, Jake stood firm. They agreed to leave the matter until the next day. The following morning, Sam Ketcham called to say the car had to come with the company. Relieved, Jake declined and the deal was off.

Sam was typical of the owners of Cariboo lumber companies among whom Jake found

Above: Completing the new beehive burner at Williams Lake, Summer 1974.

Left: Williams Lake mill in the early 1970s.

himself. They were a friendly, straight-shooting collection of the toughest competitors one could ever hope to encounter. Without a drop of malice, and without the slightest hesitation, any of them would have swallowed Lignum in a moment, given half a chance. That part was simply business; on a personal level, however, the Kerrs were welcomed into their ranks. Sam, along with most other owners, became a close friend of them both.

Advice from a few senior advisors, especially Sidney Eger and Edgar Grossman, made survival in these shark-infested waters possible. Still lacking at this point was the kind of visionary general manager required in the production division. At the time of Leslie's death, Reid Banford, an old-style manager from Kamloops, had just taken over the job of mill manager. He was unpopular among employees and for a variety of reasons it soon became clear the first priority was to replace him.

At this time, an unsolicited query from a headhunter landed on Jake's desk providing information on a US mill man with a sterling reputation as a builder and manager of sawmills. He was also a forester. By sheer coincidence, Jake attended a sawmill clinic in Portland, where he was impressed by one of the speakers. Returning to Vancouver and comparing notes, he realized the speaker was the same person pitched by the headhunter. Jake visited him at his mill in Washington, and the upshot of it was that John Ailport agreed to visit Williams Lake and look over the operation.

Ailport arrived during the Stampede the following year and checked into the venerable Chilcotin Inn with Jake. It was a normal Stampede night at the Chilcotin: the visitor was treated to a fight in the room adjoining his, challenged to a fight in the hallway, and kept awake for most

John Ailport (right), before he came to Lignum, had earned a reputation as one of the most respected mill managers in North America.

of the night by the ongoing celebrations. The next morning at breakfast he suggested he should head back to the peace and quiet of the US, but Jake managed to talk him out of it and in the end he agreed to take the job of general manager.

Ailport filled the vacuum that had long existed at Lignum. He was the kind of person Leslie had been seeking for years. He was something of a mechanical genius, who knew from first-hand experience how to put together an efficient, modern sawmill. He had built several of the top-performing US mills. He understood the importance of matching a mill to its timber supply. And he had a great ability to work with people, to inspire and bring out the best in them.

A key factor in Ailport's success in Williams Lake was the working relationship he and John Marritt formed. Marritt thoroughly understood the Lignum timber supply and, once he conveyed that information to Ailport, the new manager was able to design and build the right mill.

From working closely with Leslie, Marritt had become familiar with the intricacies of BC timber politics—a subject about which Ailport knew nothing. Between them, and with Sidney Eger's assistance in Vancouver, they were able to tie down a future timber supply that fit the needs of the new mill. They were a formidable team.

As manager, Ailport quickly realized that a key company asset was its long-term employees who enthusiastically acquired the skills and experience needed to make the new mill operation work efficiently. He set up a Ten-Year Club to acknowledge their value and their contributions.

He was a gregarious guy who, in spite of his initial experience during Stampede week, fit easily into Williams Lake, which, by this time had a population of almost 5,000 people and had acquired the status of a town. It was on the verge of a decade of tremendous growth into a community in which Ailport would thrive, and to which he would contribute his considerable energy.

A World War II navy veteran, an avid hunter with a gruff manner and a huge stuffed moose head hanging above his desk, Ailport was tough as nails when it was required. But he also had a soft heart and a real empathy for the people with whom worked. The employees liked him from day one, and the union respected him because he knew what he was talking about and could spot a bluff from a mile away.

Ailport was also superb at selecting new employees. One of the first hiring issues he had to deal with was a proposal by the mill's personnel manager, Jo DeMarco, to employ women in the mill. Like most Interior sawmills, the Lignum operation was a male bastion, and it took a lot of discussion before Lana Squinas, the company's first female mill employee, began working on the planer. Within a few months, 11 women were working in the mill.

Jake and Ailport put together a three-year capital program worth almost $3 million. The upgrading program begun then has never ceased. At about that time, with a lot of encouragement from Grossman and Ailport, Lignum instituted a policy of annual reinvestment of depreciation in a program of constant upgrading.

Most sawmills go through major rebuilds of key components every five to ten years. The problem with this strategy is that the money is not always available when the refit is needed. More

> *Finding John Ailport was a total fluke, but it was a monumental moment in the life of the company. I never cared about getting bigger. I started out big, if you will. We had a bunch of operations and I found no joy at all in scrambling around trying to figure out how to get enough money to do one thing or the other. I wanted to get one place that really worked, and I was really pleased when we started to be able to do that.*
>
> *It was also clear to me that the biggest challenge we had was getting enough money to Ailport to be able to do the things that had to be done. The partnership between him and me worked real well because I didn't make any bones about the fact that I knew dick all about a sawmill but I did know a little about marketing and I did know a little about how to raise money for him.*
>
> JAKE KERR

Some of the first women hired at the Williams Lake mill. (L to R) Susan Walters, Phyllis Proulx, Ruth Kellar, Diane Desfosses, Marion Payne, Jean Deline, and Lana Squinas.

importantly, a major refit requires a lengthy mill closure, which disrupts cash flow, especially undesirable during a hot market.

Over the next 30 years, beginning in the early 1970s, the Williams Lake mill underwent constant improvement. Only rarely was production interrupted, with most refits done over regular weekend or holiday closures.

In 1973, under its second-generation owners, some decisive moves were made. The company's head office was moved to the 15th floor of a new building on Georgia Street, at Thurlow, in Vancouver.

After his return to Lignum, Tim moved to the Interior and spent most of the next eight years working in the business from the ground up—in the woods and the mills, with time at Tappen, Salmon Arm and Tatlayoko, and even more in Williams Lake. Officially, he was Lignum's vice-president, but in fact he spent a lot of his time filling in for absent workers, doing joe-jobs and generally getting his hands dirty and his boots muddy.

With one foot still firmly planted in the counterculture of the 1960s, and the other solidly focussed on sharing the load with Jake, he cut an unusual figure in the clean-cut, straight-laced corridors of the Interior forest industry. He gravitated toward the people-oriented end of the business and began to lay the groundwork for some important long-term relationships in the Cariboo community—with natives, the union and industrial associations.

As Jake grew into the role of corporate leader, becoming in some sense the company's new "brain," Tim immersed himself in its operational intricacies, becoming its "heart" or "soul." These are, perhaps, fanciful characterizations but they reflect the distinct roles the two brothers played in taking over and revitalizing the company. Those who worked with them on those days recall Jake as a leader they respected and Tim as a co-worker they liked as a friend. Together, they worked through the 1970s, first to stabilize what their father had created, and then to build upon it.

Lignum's senior managers in the Vancouver head office boardroom, early 1980s. (L to R) Sidney Eger, secretary-treasurer; Tim Kerr, vice-president; Jake Kerr, president; Wally Pierce, sales manager; Dick Dobell, manager of the export division.

Aerial view of the Williams Lake mill yard
in the mid-1970s.

By 1974, the company's new strategy began to emerge, not necessarily fully formed but in bits and pieces, some of it acquired in the school of hard knocks. The markets that year were worse than they had been in 25 years, but some major improvements went ahead at Williams Lake. The log decks were rebuilt and a new debarker and grapple installed. An edging picker, which automatically diverted slabs to the chipper, eliminated a backbreaking, mind-numbing job. The saws acquired carbide teeth. In the planer, a new chipper was installed, a strip stacker put into operation and a blower added to move shavings to the sawmill burner.

In April, Tatlayoko burned to the ground. A fire that began in the mill spread through the entire operation, burning houses and vehicles, stranding the 35 employees and their families. In the end, the fire proved to be a mixed blessing. The mill's timber supply was shrinking and there was a growing shortage of timber in the Chilko PSYU. The Forest Service was reluctant to have the mill reopen and, after heavy negotiations on the part of Jake and John Marritt, officials agreed to transfer Lignum's Chilko quota to the Quesnel PSYU.

Established operators in Quesnel did not greet this decision with enthusiasm. However, they were mostly utilizing fir and spruce, and avoiding lodgepole pine. After a careful look at the Quesnel PSYU inventory, Marritt discovered substantial volumes of high-quality pine everyone else had ignored. He and Sidney convinced the others it would be a good strategic move to go after the pine. The difficulty lay in finding an operating area with little or no fir and spruce in order to avoid conflict with the established licensees. It was a delicate situation and a milestone in Jake's education in the nuances of timber-supply politics.

Jake had received a baptism a couple of years earlier, under heavy fire. A few months after Leslie died, he made an offer on a small sawmill near Springhouse owned by Martin Gunstveit, an old family friend whose lumber Lignum sold. The deal went bad, and only with great difficulty was it eventually sorted out. But it left a lasting impression about the pitfalls of growing a company through acquisitions.

> *One of the major turning points for Lignum's forestry was the fire we had at Tatlayoko, which brought an end to that operation. Then the government transferred the quota out into the Quesnel PSYU. Of course that wasn't well received by the other licensees, especially Merrill Wagner. They were shocked. The forest service arbitrarily gave us a quota in the Quesnel PSYU. We had an interesting time finding an operating area in the Quesnel. Jacobson was already there, and Merrill Wagner, and West Fraser too, and they didn't want us there at all. Wherever we applied they said, "You can't go here, you can't go there."*
>
> *Finally, I picked Merrill Wagner's choice area and marked out a big chart area. I sent it into the government and sent Merrill Wagner a copy. That got their attention! They said "Okay, okay, maybe we should get together and find you an area we can all live with." That's what it took to get their attention and cooperation.*
>
> JOHN MARRITT

Top: Williams Lake mill's new log sorter in action.

Bottom: Constructing the base for the eight-inch Chip-N-Saw, late 1970s.

Another lesson of this sort would be learned a few years later, and develop even further Lignum's corporate strategy. This deal involved the purchase of Komori Lumber—Laci Kardos' old company—in 70 Mile House. It was a straight-up buyout of a sawmill, planer, logging operation and a good supply of timber. The mill was facing a timber shortage and eventual closure. Lignum was only interested in the timber, which it planned to use in its Williams Lake mill.

The Ministry of Forests and the politicians involved—forest minister Tom Waterland and local MLA Alex Fraser, who was minister of highways—agreed to the sale, with full knowledge the mill would be closed. An hour before signing the sales agreement, Jake confirmed the ministers' concurrence, even though no formal permission for the sale was required.

When the sale closed, the people of 70 Mile House objected vigorously, demanding action from their MLA, Fraser. Ainsworth Lumber entered the debate, offering to buy Komori and keep the mill open. The politicians ran for cover, threatening to prohibit transfer of the timber licence and to prevent log hauling to Williams Lake on the highway.

Accompanied by Peter Stanley—Edgar Grossman's newly acquired junior partner—Jake obtained a meeting with Waterland and Fraser in Victoria. Pointing out the lack of any legal basis for the government threats, the Lignum contingent conveyed their intention to initiate personal lawsuits against the cabinet ministers. That settled the matter, then and there. The sale proceeded, and a few months later the government passed an amendment to the Forest Act requiring the minister's approval for timber transfers.

The final piece of the new strategy fell into place in 1975 when the Salmon Arm and Tappen mills were sold. That same year, the markets improved. With the mill improvements and Ailport's changes in effect at Williams Lake, the company started showing some good profits. The idea of selling it began to fade, as the new business strategy took hold.

With the sale of the outlying mills came the realization that, given the management structure of the company, Lignum could not effectively run more than one operation, especially if

We looked at deals. You could argue that it was a lack of courage on my part, but I didn't want to take on a leverage, I didn't want to take on the debt because I kept watching guys that I respected get into trouble.

Some didn't. Some were excellent at it. A guy like Sam Ketcham at West Fraser, boy did he understand what he was doing. He was willing to take risks that I wasn't willing to take. The first deal that I made after my dad died, I wasn't always as focussed as I've become, with a strategy that says you stick to your knitting. We went out and bought a small sawmill right next to Fred Linde, owned by Martin Gunstveit, one-armed guy, an old Swede.

We did that deal two months after my dad died. Bought him out for $100,000, and when it came time to close, we didn't have $100,000. I had to stiff him and I felt really bad because he was a great old friend. So he sued us.

At the end of the day, Mr. Grossman said, "You've got to find $100,000." So as it turned out, we went to the bank and we paid him $100,000. It was a really unfortunate, unpleasant affair.

A few years later, we bought Komori Sawmills, down at 70 Mile. It was a brilliant acquisition, I give us full credit for it. We paid 2.2 million bucks for it. Alex Fraser, the MLA at the time, tried to block the deal and we threatened to sue Fraser and Tom Waterland. I got pilloried in the press. I just got murdered by the citizens of 70 Mile for closing the sawmill. You couldn't do that today. Lignum's purchase of Komori caused them to amend the Forest Act so you couldn't do a deal without ministerial approval. I didn't need it at the time.

So you wonder why I didn't do more acquisitions? I did two real quick and I got my ass kicked all the way around the block! I also could see, once we got cooking, that we were making real strides. You think about it another way: Why would I have wanted to acquire another sawmill when it seemed we couldn't run the one we had.

JAKE KERR

Lignum sales managers inspect leased Thrall lumber cars, acquired to alleviate a PGE car shortage, June, 1972.

additional mills were any distance from Williams Lake. Nor were the capital funds easily available, without taking on high levels of debt or entering into a partnership to finance such expansion. In later years consideration would be given to the purchase of Merrill Wagner and Pinette and Therrien, when they came up for sale. But, as they arose, these options were rejected and through the latter half of the 1970s, the idea of focussing on one mill complex and making it as efficient and profitable as possible began to gel.

Part of the process involved the computerization of the company, which had started with the purchase of the first computer, actually a glorified bookkeeping machine, in 1970. By 1975, when Lorraine Pungente was hired to cope with the new technology, a computer department existed. The following year, the Williams Lake and Vancouver offices were connected via a telephone line for transmission of payroll data. After that, it was mostly a matter of keeping up with the rapid changes in hardware and software, a program of steady improvements similar in principle to that followed at the mill.

In 1975, the government established another Royal Commission on forestry, conducted by one of Jake's former university professors, Peter Pearse. Jake, Tim and John Marritt presented Lignum's brief. It was exactly that, brief and to the point. The primary recommendation was for a longer-term form of tenure for harvesting licences.

In the mid-1970s, a contentious issue within the Interior forest industry, chip prices, came to the fore. When the provincial government reorganized the Interior forest industry to accommodate a pulp sector, sawmills such as Lignum's were required to ship their chips to specific mills—in Lignum's case, to the Canfor pulp mill in Prince George. The way this arrangement evolved, the pulp mills ended up setting the price of chips. By the early 1970s, this price was unrealistically low, less than $10 a bone-dry unit—far less than what it would have cost the pulp mills

to produce chips from raw logs on their own. One factor which kept prices low were regulations that made it difficult, if not impossible, to export chips from the province.

After meeting with mill owners and reviewing the situation, the NDP's forest minister, Bob Williams, threatened to have the government set chip prices, and began to relax the conditions under which chips could be exported. While most sawmill operators preferred a negotiated solution to the problem over government intervention, the price increases for chips that resulted were much appreciated. At that time, a huge surplus of chips existed throughout the Interior in mountainous piles towering over virtually every sawmill.

After the Social Credit party came back into power in late 1975, it loosened export regulations even further, enabling Interior chip-producing sawmills to negotiate long-term export contracts with buyers outside the province. Lignum made a couple of export sales on its own to Japan, as did a few other independent sawmill companies. Soon after, several mills in the central Interior, including Lignum, collaborated in the formation of a chip-export company, Fibreco, with a reload dock for export shipments in North Vancouver. The ability to export chips when surplus supplies build has kept prices at more reasonable levels ever since.

The company was beginning to come together as a coherent, effective entity. The markets stayed strong, profits increased and funded improvements. This was the moment the sales staff chose to stage a mutiny.

Lignum was selling lumber for several Interior companies, including the output of Lavington Planers, run by John and Al Thorlakson out of Vernon. Like other mills Lignum sold for, the Thorlaksons had a desk in the Lignum office, to keep an eye on the markets and learn something about lumber sales. They got to know the sales staff well.

With the company's bottom line looking better every year, and dissatisfied with their share, Ken Ross and most of the sales staff quit, striking a deal with the Thorlaksons to set up a sales department for the Okanagan company in Vancouver. At this time, the Thorlaksons' company was renamed Tolko. Instead of the usual practice of working as employees on salary to sell the mill's lumber, the sales group obtained a profit-sharing arrangement.

The mutineers were joined by the Luckhurst brothers, Lee and Mick, a couple of high-powered lumber traders whose family owned a piece of Raven Lumber in Campbell River. They had a great time, and made piles of money for themselves and for Tolko.

Eventually the Thorlakson brothers, who ran a pretty staid operation, decided a bunch of high-flying Vancouver lumber traders working inside their business was an unwholesome influence. They terminated the sales agreement and set up their own, tightly-controlled sales office in the Okanagan where they could keep a close eye on it.

Meanwhile, Lignum's sales department, left in the lurch by the mass exodus, struggled to get back on its feet. The only sales staff who remained were Wally Pierce and Jim Gray, who scrambled to hire new staff and keep lumber moving out of the mill yard. The wholesale end of the business dropped from about 40 percent of sales to about 10 percent as the sales department focussed on selling Lignum's own production.

After leaving Tolko, Ken Ross and his crew continued on their own for a year or so and then ended up back at Lignum, along with the Luckhurst boys. Lee Luckhurst was made operations manager of the division.

By the time of the mutiny, in the fall of 1977, Lignum's production division was solidly on its feet. Under Ailport, lumber production was consolidated in Williams Lake and had become a highly efficient operation.

A practice implemented by Ailport and Tim was

Two parts to the business that made sense to the type of mind my dad had were brokerage and forestry. He was very, very wise at figuring out that the timber positions were what was worth the money. His worst thing is that he could never figure out how to run the sawmill. Until Jake and John Ailport took over in '72 and we started to be able to reduce our costs—West Fraser and the other guys had a 10-year head start on us in low-cost lumber production.

My father, I think because of a lack of capital and a lack of really understanding sawmills, never had guys with expertise about manufacturing. The first real mechanical wizard we had was John Ailport. He was like Ken Siemen up at Quesnel with West Fraser, Doug Floyd and other guys. They knew how to make sawmills really run. That changed Lignum completely because we had stumbled around selling the stuff quite well, and ran the forestry end in an aggressive way. And then running the logs through a mill operation that didn't run that well. Ailport changed that. I guess Ailport actively managed the mill for 10 or 12 years and he taught a bunch of people how to manufacture lumber.

~

TIM KERR

Gian Sandhu at a company picnic.

to encourage and underwrite high levels of training for employees. It was part of a broader policy of recognizing good employees, encouraging them to develop their skills, and rewarding them for their accomplishments. This was a new way of thinking in the forest industry which, until that time, had largely fed off the bottom of the labour pool, employing a lot of untrained, marginally skilled, transient workers. Turnover rates were high, but since it did not take a brain surgeon to pull lumber off the greenchain in an old-fashioned mill, filling vacancies was rarely a problem.

The extremely sophisticated mill Ailport began creating during the 1970s could not function with this kind of a work force. Nor were the good people who were available much interested in the labour-intensive bull work offered in the old-style operations. The most desirable employees wanted steady jobs that rewarded them for good work in order to raise families in the attractive community Williams Lake was becoming.

Gian Sandhu was a perfect example of the kind of people who came into Lignum at that time and helped it achieve the success it experienced during this period. A former officer in the

When we moved into pine the hauls got longer and longer, and we realized we were hauling a lot of wood that really didn't have any value. A whole tree there is only 35 feet, 40 feet, and certainly not 50 feet. If we got a 50-foot tree we thought we were doing pretty good. So the quality out there was very questionable to start with, very tough.

But that's how we started and we did that for many, many years, until the logging equipment developed to the point that we looked into short-log logging. We found that a large part of the wood was of poor quality. We were better off to buck it up into short logs and make sure that every log we hauled long distances was of value to the sawmill.

If you talk to a mill manager, Christ, they claimed they delimbed every tree we brought in! In my whole history with Lignum, I never did bring in one good log, from their point of view.

I understand that, I've had to live with it all these years, but we always delimbed out in the woods. Eventually we went to

the next stage where we started to buck into short logs in the woods. That way we could ensure that every log we brought in was a good saw-log for the mill.

We mechanized bucking in the late '70s. We did a number of tests in the mill before we did that. We'd take some of our pine loads in the log yard and buck them up into shorts. We got quite an inventory of nice short logs and fed the mill that way. We found the mill just went zoom! It just sucked them up like nobody's business and made great lumber out of them. With that encouragement, we said, "Hey, maybe this is what we should mechanize out in the woods. So we went through that phase, which needed a different configuration for our logging trucks and bunks. Some of them who had to really modify their trucks, we'd help out with advances and things like that. It was a very trusting relationship. It was great—seemed easy in those days!

JOHN MARRITT

Indian air force, he came to Williams Lake in 1971 shortly after immigrating to Canada. He worked at Lignum for the next 10 years, starting at a menial job and quickly working his way into more responsible positions, ending up as purchasing manager. With Sidney's assistance, he obtained an MBA in his spare time.

Coincidentally, the year Leslie Kerr died, Dean Getz fell ill and his son Roger took over control of three-quarters of San Jose Logging, along with his uncle Joe. Encountering estate problems, Roger made a similar decision to the one that brought Jake back into the company— get it into good shape and sell it. In pursuit of this goal, by the end of the decade he had created one of the largest and most productive contract logging operations in the Interior. And, like Jake, having succeeded at his initial task, Roger stayed on. San Jose has remained as Lignum's primary logging contractor ever since.

San Jose Logging's grapple skidder (Top) bringing in a bunch of whole trees to the landing, where they are delimbed and topped before loading (Bottom) for trucking to the mill. A typical logging system of the 1980s.

*The new mill office
under construction at
Williams Lake.*

In 1979, Barbara officially opened a new office at the Williams Lake mill. From the day Lignum moved to its permanent site the office staff had made do with a 500-square-foot building that acquired occasional additions. After its replacement was built, Jake took great delight in taking the controls of a bulldozer and flattening it. A new office had been on the agenda for some time, and in their 1973 wish-list, Jake and Ailport budgeted $125,000 for a new building.

The one eventually built was a far cry from the original plan. Designed by Vancouver architect Peter Cardew, it was a stunning $1.5 million, 5,000-square-foot facility, which more than 20 years later still astonishes first-time visitors. The building won several architectural awards. Its basic design concept is that of an ultra-modern building incorporating structural elements of an old bush mill.

It was built with exposed laminated wood beams assembled from lumber cut at the mill, a lot of windows and industrial steel girders, stairs and railings. At the opening Cardew explained that he had tried to

create an ultra-modern building in an industrial style that was not out of place in the mill yard. A large mural composed of stylized photographs of Lignum operations was later installed in the foyer.

The office was a bold statement—"We've arrived in style and we're here for the long haul!"—which was as much a personal statement by Jake and Tim to the employees as it was a corporate statement to the community and the region. As the next 20 years would show, the office building symbolized a new beginning for Lignum.

CHAPTER 6

LIVING THE VISION

The reconstruction of Lignum and its relaunch under a new management strategy in the 1970s was fortuitous. Along with the rest of the industry the company was about to embark on two turbulent decades of rapid, fundamental change that would eliminate scores of lumber producers, large and small.

By the end of this period, at the beginning of a new century, Lignum would emerge as one of the strongest forest companies in North America. It would get there through its own, unique set of principals and practices—all of which would be seriously tested and, on occasion, found wanting.

During the 1980s and 1990s a complex series of circumstances arose that recast the context in which Lignum operated. These were decades of incredible market volatility, complicated by a running dispute with competitors in the British Columbia forest industry's major market, the United States. The legal foundations of the forest resource base were challenged by aboriginal people, creating an atmosphere of uncertainty that undermined investor confidence in the future. A fundamental shift in public perceptions about forests and how they are used occurred during this period. And, finally, a series of enormous insect attacks on forests in the central Interior forced sweeping changes in the management and utilization of the industrial timber supply.

The 1980s began with the virtual collapse of forest-product markets in what quickly became the worst global economic recession since the 1930s. By this time, most of the lumber produced by Lignum and other Interior mills was going into the US.

When this market was strong, the US domestic lumber industry was incapable of meeting demand. Under these conditions Canadian lumber filled the gap and profits were high. But with a recession, the market for new houses in the US fell off dramatically, the demand for lumber weakened, and prices plunged.

Given the fundamental differences in the Canadian and US forest industries, responses to the downturn differed in the two countries. US timber owners stopped logging their private lands when log prices fell below a certain point, and curtailed production of lumber.

Lignum maintained production through the recession of the 1980s.

In BC, sawmills continued to operate, even when lumber sold at a loss. The dominant timber owner, the provincial government, encouraged continued production in order to maintain government revenues and employment levels. In the early stages of the downturn, pulp markets remained relatively strong. The demand for chips produced as a lumber by-product in Interior mills—almost all of which by now were equipped with Chip-N-Saws—remained strong. The mills continued cutting lumber to generate chips. The result was a surplus of lumber flowing into the US from BC, depressing prices even further. Some US forest companies objected to this situation and in 1982, under provisions of US trade law, they initiated a countervail duty action against BC lumber.

The BC Council of Forest Industries led the negotiations for industry and Jake was a member of the negotiating team. The US industry lost this dispute. Jake's experience studying and working in the US, supported by Edgar Grossman's knowledge of American trade policies, made him a valuable member of the team. It also gave him important insights into the working of US lumber markets.

The obvious solution to the tightening of the US markets, especially for a company such as Lignum with its own sales division, was to find buyers outside the US. It was a deceptively simple solution, and one Leslie had approached cautiously. He had been badly burned on at least one occasion when a big order to Britain went bad. After that, his primary strategy, the importance of which Jake and Tim had absorbed around the family dinner table, was to devote a great deal of effort to maintaining unrestricted access to the US market.

The trading floor in the Vancouver office.

It was one thing for a group of traders working in an office like Lignum Sales to operate effectively in the North American lumber market, where most of the players were known. In other parts of the world, lumber markets function according to other rules and protocols. There are many players, all working to different standards. If a trader does not have experience in the global market, expensive mistakes can be made.

Well aware of these risks, Lignum hired a couple of traders with extensive experience in off-shore lumber sales at Seaboard Lumber Sales to set up an export division. Dick Dobell, who was raised in Britain and whose family was well-connected in the lumber trade there, headed the export division and brought Ron Ogilvie, another Seaboard stalwart, with him. Both more than six feet tall, the two were known as The Giants. They proceeded carefully, developing markets for high value products, initially in Britain and Japan, then in Europe. Within two years, Lignum was selling between three and five million feet a month into these markets. A significant portion of sales was through an English intermediary, Maurice Eaton, through whom Lignum sold lumber for a decade or more. To a certain extent, these sales took Lignum back to its origins, when Leslie relied on Europe for a substantial portion of the company's business.

Even so, it was easy to overlook all these factors in 1981 when Rafiq Sharif, a previously unknown Egyptian buyer, approached Lignum in what Jake described at the time as "an amazing cold call" to order 10.5 million board feet of lumber worth $6 million. Best of all, he wanted the

Left: The Egyptian adventure. (Left to Right) Dick Dobell, Vic Nightscales, Rafiq Sharif, and an unidentified Sharif employee in front of unloaded Lignum lumber on the Alexandria docks, 1981.

Above: Lignum managers Dick Dobell and Vic Nightscales on the job in Egypt.

low-end utility and economy grades that were most difficult to sell in the falling North American market. Sharif's credentials were confirmed by his Cairo bank and the first-ever full-cargo shipment of lumber from BC to Egypt was underway.

Lignum assembled the order from its own and a half-dozen other Interior mills, and dispatched the first of three shiploads to Egypt. When it arrived, the longshoremen refused to unload it. Then other "problems" arose. The buyer, it turned out, was a shyster. He was a principal in the bank that vouched for him, and he controlled the longshore business.

The basic problem, however, was that because the company was untutored in Mediterranean lumber trading protocols, Lignum had neglected to pay the required "fees" to permit unloading. Dick Dobell and another member of the executive, Vic Nightscales, flew to Alexandria to grease the requisite palms, and the shipment was unloaded.

Fortunately, Lignum collected on the letter of credit it had received, but it later blew up in the Bank of BC's face. By this time it was obvious this was no place for a nice bunch of BC boys to be doing business, so they terminated the transaction. Unfortunately, Lignum had bought all the lumber for the second shipment, which was already in transit from the Interior to the dock in Vancouver when the deal fell apart. It consisted of a lot of oddball sizes and grades and, even with Maurice Eaton's best efforts, disposing of it was a nightmare that ate up all the profits from the first ship load. It was an expensive learning experience in the Egyptian lumber business.

Beginning in the late 1970s, Cariboo-Chilcotin forests came under heavy attack by mountain pine beetles, creating enormous problems for the forest industry. These beetles are a natural phenomenon in lodgepole pine forests, normally existing in small, scattered populations. Under the right conditions, their numbers can increase astronomically. Pine beetles prefer older trees and, because of successful fire-fighting efforts by government and industry, and limited timber harvesting, Interior pine forests have generally increased in age to become susceptible to large outbreaks when climatic conditions are favourable. The insects thrive during dry summers and warm winters, both of which occurred with increased frequency after the late 1970s.

During the early 1980s, a major pine beetle attack occurred in the region, and continued until 1985, leaving behind in Lignum's operating area 600,000 hectares interlaced with patches of dead trees. Lignum had begun running logs from bug-killed trees through its mill in 1979 and

Mountain pine beetles infested the Chilcotin forests in the early 1980s.

found they produced acceptable lumber. At the height of the infestation the government made beetle-killed wood available at special low stumpage rates as an incentive for its salvage and use. At the time, it was uncertain how long the dead trees would remain standing and free of decay.

The initial experience with bug-killed wood convinced Marritt and Ailport it had potential. In 1981, Ailport was made vice-president of facilities and design and Marritt succeeded him as general manager, giving them increased latitude to innovate.

For the previous three or four years, Marritt had been working closely with Lignum's logging contractors to develop more efficient methods of harvesting and delivering small Chilcotin pine logs. He was also watching closely the beetle infestation, and the dead timber left in its wake. He had a pretty good idea how much was available and what it would cost to harvest.

In the mill, Ailport was discovering that the dead pine trees, which had dried standing, produced stable lumber that did not warp or twist after it was milled. Coming from the extreme climatic conditions of the Chilcotin plateau, lumber milled from the small, slow-growing trees was stronger than that from faster growing trees on more productive sites. These factors made the dead trees particularly suitable for some specialized uses.

At about the same time, through his contacts in the US, Jake became aware of a rapidly growing market for stress-rated lumber used to build roof and floor trusses. The machinery used to determine the strength of lumber was just becoming available.

In 1985, with markets showing an upward trend,

Lignum's MSR lumber, manufactured from beetle-killed trees, used in truss construction in the southwestern US.

Lignum's first carload of two-by-four MSR ready for shipment from Williams Lake, August 13, 1986.

it was decided to begin producing machine stress-rated (MSR) lumber from the beetle-killed pine for sale to truss makers in the US. This was a major decision, involving a $4 million reconfiguration of the mill and a substantial retooling of woods operations. The dead trees were scattered, usually in small patches, often of only a few trees. The trees, typically, were small and often malformed. Logging them required contractors to develop new logging systems and purchase new equipment, and share with Lignum the risks of an uncertain supply, not to mention the untested markets for its products. Logging costs increased by about $15 a cubic metre, which was only partially offset by lower stumpage rates. Lignum was one of the few companies in the region to refocus extensively on this wood.

In the midst of this conversion process, several important changes in the company's management structure occurred. In 1985, Barbara died. She had served as board chair since Leslie's death, a role now taken on by Jake. Tim became president.

At roughly the same time, Ailport retired. He did not go far, staying in Williams Lake, where he was involved in various community activities, and coming into Lignum regularly to put out the staff newsletter. Dwight O'Donnell, a brilliant mill designer and operator from Tacoma, came in as mill manager and oversaw installation of the MSR equipment. Marritt refocused on the woods end of the operation for a few months, before moving to the Vancouver office to work on corporate planning and forest policy development. Phill Nelson took over as woodlands manager.

Stress testing equipment was installed in the mill to rate the strength of every board and label it with a grade stamp. The first production runs took place in the summer of 1986. The sales division began developing a marketing campaign and the first sale closed in August—an order of two-by-fours that sold for a $38 premium over Standard and Better two-by-fours.

Jake Kerr and John Ailport at the dedication of Ailport Park in Williams Lake, June 1988.

The move to MSR was enormously successful, placing Lignum well in front of its competition in the utilization of bug-killed wood, as well as in the production of this specialty lumber. By the early 1990s, Lignum was harvesting 200,000 cubic metres of dead pine annually, and 50 percent of its mill output was MSR.

The addition of MSR capability required a lot of other changes in the mill which, combined with other improvements, resulted in an extensive rebuild during the mid-to late 1980s. In 1985, even before the MSR installation, the mill's beehive burner was replaced with an energy plant that burned mill waste to heat oil for use in the dry kilns—a much-appreciated decision by both Williams Lake residents, who disliked the pollution the beehive burner created, and mill workers, whose jobs it threatened.

A couple of years earlier, on a dry afternoon in late May, a spark from the burner ignited a fire in the nearby log yard. It was a spectacular fire, providing residents of Williams Lake with hours of free entertainment as a fleet of water bombers dived into the billowing smoke to extinguish it. Hundreds of vehicles

All the mills, in the beginning, were pretty well convinced that cut-to-length logging was too expensive to work. It may cost a couple of bucks a metre more, loaded on the truck, but if you look at the benefits of the thing, it's actually costing less because you build less roads, you don't have to do any follow-up burning, that type of thing. It may have silviculture benefits as well. All the debris is in the bush. Wetter areas, it's not a big deal because a lot of it rots very quickly.

You can operate pretty well year 'round, because you operate on the mat, in other words the harvester processes and crawls along on top of the debris, the limbs and so on, and the forwarder comes in behind.

What they're looking at is when we're transporting this timber from 200 miles out, the shortwood, we can get a lot more solid wood on a load. It cuts down on the waste hauled. All long logs, there could be rot, there could be all kinds of things in those logs that you don't know about, and you're paying for that. So there's cost savings in the trucking.

We're fortunate, we're working for probably the most progressive mill in the province, if not North America. They've been aeons ahead of everyone else in terms of developing these new systems.

Dave Rankin

Water bombers attacking a fire at the Lignum mill, 1983.

and many more people gathered on the hillside above the mill to enjoy the spectacle. Mostly by chance, the mill escaped unscathed, but more than $100,000 worth of logs were destroyed.

In addition to the MSR equipment, a mechanized sorting system—a J-bar sorter—was installed in the planer mill, and a new planer added. A new, $3.7 million edging system was added to the sawmill, along with a host of other improvements. By the end of the decade, under O'Donnell's guidance, Lignum's Williams Lake mill was one of the most sophisticated in North America.

The mid-1980s was a pivotal time for the company's mill and woods operations. It was the end of the Ailport-Marritt era, and the beginning of a new one fraught with uncertainties and change. It was also the beginning of a new approach to management, and the beginning of a shift in emphasis away from old-style lumber trading to a new, customer-driven approach.

Mulvany and Associates, a US management consultant firm, was hired in 1983 to reformulate the company's management practices. It was particularly effective in the mill operations where, as in every other sawmill in the country, a conventional hierarchical organization prevailed. By the early 1980s, a situation had evolved where communications between mill workers and managers had settled into a pattern. Employees were told when they did something wrong, with little said when they did a good job. A lot of grievances were filed with the IWA and a lot of work stoppages occurred.

Jake and Tim met Steve Mulvany, an up-and-coming management consultant, and invited him to implement his program, called "Management Tools," at Lignum. The program was built on two assumptions: one, that people want feedback, and, two, that they want to excel. First, the program determined how information flowed within the company, and suggested what information was needed by employees to do their job better. Second, it looked at the way people were supervised, and recommended a new style of management.

Lignum staff at a Management Tools workshop, 1980s.

Although the Mulvany program was applied to the entire company, its biggest impact was felt at the mill. Jake, Tim and other senior managers participated in all of the extensive training programs conducted over the course of an entire year, under the supervision of personnel manager Mike Skellett. Initially, the program was received with suspicion and, in some cases, distrust. The union publicly opposed it as a union-busting effort conducted by American consultants to weed out less productive workers.

Eventually, most employees came to appreciate the benefits. They were able to perform their jobs better, and receive the rewards for their improved performance. Good work was not taken for granted, but acknowledged and rewarded. In the end, the union came around and the program has become a permanent part of the mill's operations.

Following the return of the mutinous lumber traders in early 1982, Lignum's sales division increased its wholesale activity, reducing mill sales to about 60 percent of the much-increased total sales. Under Lee Luckhurst an aggressive wholesale campaign was mounted. The sales staff tripled, the export division was relaunched and a number of other initiatives undertaken. One of these, even before the mill began MSR production, was the establishment of an MSR marketing office for which Aaron Anderson, an experienced sales manager of the product at Weyerhaeuser, was hired in 1986.

The mill yard with a full inventory of logs on hand, mid-1980s.

Encouraged by its growing success, the sales division also opened an office in Toronto, known as Lignum East. Its task was to wholesale truckload quantities of lumber and wafer board from reload centres in Windsor and Niagara Falls to markets in the northeastern and mid-western US. It also sold carload lots into eastern Canada. Dan Joerges and Arsen Zezelic, two veteran Ontario traders, were hired to run the operation.

Shortly after the office opened in early 1988, the market dipped and within months it was losing significant amounts of money. With memories of red-ink divisions such as the Tappen and Salmon Arms mills still fresh, Lignum pulled the plug on the eastern office before things got worse. At the end of the year, Lignum Sales was rolled into Lignum Ltd.

Closing the eastern office was the beginning of a general retrenchment. Even before the office was opened, Dick Dobell and Ron Ogilvie were let go and the export effort scaled down, essentially to the British business going through Maurice Eaton. Convinced, perhaps, that the growing wholesale business would continue to thrive, the Vancouver head office and sales offices had been moved to the top floor of a luxurious new building on Burrard Street, Park Place, in 1985.

In the wake of the Lignum East disaster, the Park Place offices were sold for smaller quarters on another floor of the previous building at 1090 West Georgia Street. Some of the high-end furnishings were brought over from Park Place so that—given the company's art collection, which was probably the best corporate collection in the city—they were among the most elegant offices in Vancouver's downtown business district.

At about the same time, a purge was executed in the sales department. With the return of the mutineers and the ramping up of the wholesale end of the sales division, a freewheeling style of trading had predominated. A lot of lumber was sold and a few key traders earned small fortunes in commissions. But the usual conflict between the interests of the individual traders and the company's overall interests increased.

Ken Ross had left Lignum even before the eastern fiasco, which was Lee Luckhurst's initiative.

Following the closing of the eastern office, Lee and several of his closest colleagues were terminated in 1989. Aaron Anderson, who was now heavily engaged in marketing MSR, was put in charge of the entire sales division, and additional MSR sales staff hired.

Selling lumber into the US at this time was further complicated by a resumption of the trade dispute with the US lumber industry, which itself was complicated by a significant shift in BC politics. Jake, in his capacity as Lignum's CEO, was thrust into the midst of these events.

During the late 1970s and early 1980s, he had served as president of the Cariboo Lumber Manufacturers Association—successor to the association Leslie served as its first chairman—where competitors within the region hammered out their differences and forged alliances. Following his involvement in the 1982 trade dispute, Jake was elected chairman of the Council of Forest Industries, an umbrella organization representing most of the BC forest industry. At this time, 1985, COFI was by far the most powerful business organization in the province and represented the forest industry to government. During Jake's term as chairman, when provincial politics went through one of its more eccentric phases, relationships between the industry and the provincial government were often strained.

In 1986, Bill Bennett had retired as premier and leader of the governing Social Credit party, to be replaced by the charismatic Bill Vander Zalm, who often displayed a hostile attitude to the forest industry. In their first meeting with Vander Zalm, Jake and COFI president Mike Apsey were surprised to discover the new premier had no knowledge of, and very little interest in, the forest sector. His primary concern at the meeting, which took place at his amusement park—Fantasy Gardens—was to pose for photographs with visiting tourists.

Vander Zalm came into power shortly after the US lumber industry initiated the tariff action. Jake was heavily involved in this round of negotiations, in the thick of which, Vander Zalm and his erratic forest minister, Jack Kempf, intervened in a manner that, essentially, strengthened the US position. The end result was the imposition of a federal export tax on Canadian lumber shipments to the US, which the following year was replaced with increased stumpage fees and a new stumpage system in BC. From this point on, lumber trade with the US was complicated by the involvement of politicians.

Lignum concluded the decade, which had seen considerable internal change, with a reformulation of its senior management structure. By this time, Jake's involvement in the industry extended far beyond the operations of the company. It was apparent that fundamental shifts in provincial, national and even global forest policies were well underway. Changes that could be foreseen only dimly a decade earlier were now in full flower. Not only was Jake involved in the development of these policies, but the outcome of the growing debates would clearly have a significant effect on the future of Lignum.

At the same time, Tim wanted to withdraw from daily participation in company affairs to return to university and to spend more time involved in the education of his three young sons. Coincidentally, Sidney Eger retired, although he continued to come into the office and serve in an advisory capacity for the next decade and a half.

Ron Longstaffe, a former Canfor senior executive, was hired in 1988 as Lignum's executive vice-president to take over some of Sidney's duties and to give Jake more time to work on industry affairs. As had usually been the case in previous appointments of managers from bigger companies, this one too proved difficult. Large public companies typically develop substantial administrative bureaucracies and procedures. In a company like Lignum, where the owners occupy management positions, committees do not exist, and decisions can be made quickly and informally. Longstaffe left after about one year.

Late in 1989, Jake held a series of discussions with Conrad Pinette. Conrad had concluded the sale of his family firm to the successor of BC Forest Products, Fletcher Challenge, a New Zealand firm run by money managers more concerned with extracting assets than building viable long-term operations. This was not Conrad's kind of place so, after lengthy discussions about the Kerrs' plans for the future of Lignum, he accepted an offer to take on the job as its president and chief operating officer.

John Thomas, a career banker, came into the company as vice-president of finance at about the same time. Aaron Anderson was already installed as the new sales manager. At Williams Lake, Dwight O'Donnell left to return to the US and was replaced by Rob Fraser, who had started at the mill in 1981 and worked his way up through the ranks. He was its first home-grown general manager. Jake took the title of chairman and CEO, and Tim became vice-chairman. A new decade was about to open with an entirely new management team.

The gang saw outfeed in the late 1970s (Top) and the same location in the mill 20 years later (Bottom).

In general, the early 1990s were not good years for the BC forest industry. Lumber markets fell off drastically from the healthy levels they reached after recovering from the recession of the early 1980s. In 1990, the province's sawmills experienced combined losses of almost $300 million. Not all these losses were due to market factors, however. The industry was encountering structural deficiencies that would increasingly plague it into the future.

Environmental organizations, particularly on the Coast, were growing in size and influence. The forest industry was increasingly portrayed in the media as an out-of-control, destructive force in the province's forests. The Vander Zalm government was degenerating into chaos, and in 1991 the quasi-socialist New Democratic Party under the leadership of Mike Harcourt swept into power. That same year, the US lumber industry precipitated yet another tariff dispute.

In retrospect, it is clear Lignum was in better shape to cope with the coming difficulties than many other forest companies in the province. Its mill was perhaps the most efficient in existence, with an ongoing program of capital investment that made it possible to maintain that lead. Its move to MSR placed it two years or more ahead of most potential competitors in this specialized product niche. A new and much younger sales force was gearing up to take advantage of opportunities in this market. A strong, new management team was in place, leaving Lignum's chairman free to focus on external matters vital to the company's future.

We evolved over 10 years, probably, by giving the customers what they want, on time. They think of us when they are going to buy. Now all companies say that, and all companies have a mission statement, and all companies tell you that they're customer driven. But we really are, and our profit would probably reflect that. Our mill has done a terrific job with pretty poor timber, to get really good product out of that poor timber and still make money. We specialized in what we did, and we've done very well.

JIM GRAY

Resumption of the softwood lumber dispute with the US industry in 1991 precipitated a struggle that continued almost unabated until early 1996, when a five-year Softwood Lumber Agreement was signed. For the duration of this episode, Jake was centrally involved in negotiations, the last two years as chief negotiator for the Canadian industry.

Given the nature of Lignum's business and the evolution of the industry as a whole, it was a curious and at times delicate position he occupied. Much of his time was spent in BC and Canada, developing positions and strategies with the heads of other companies, all of them competitors and most of them much larger in size than Lignum. By the late 1990s, mostly because of consolidation and a decline in the number of large timber licensees, Lignum became the 21st largest holder of Crown timber rights, with almost 600,000 cubic metres of annual cut under licence.

> *The old man told me one thing and demonstrated it: that nothing else matters but being able to ship your product without restriction to the US. So I grew up not only hearing that, but believing it, and to this day that's what I'm working on. The rest of it we can fix. Without that, we've got no market.*
>
> JAKE KERR

The 1996 agreement established a limit to the amount of Canadian lumber that could be shipped to the US duty free, which resulted in quotas for individual companies in four provinces based on their historic US sales. At the time, this agreement was widely viewed as a victory for Canadian lumber producers.

In BC, most of the quota was assigned to Interior lumber producers because coastal mills had sold the bulk of their production in Japan. Shortly after the agreement was concluded, however, the Japanese market collapsed and BC coastal mills, lacking quota in the US market, were prevented from selling into it. Similarly, eastern Canadian mills, which were at the time

Logging in the 1980s. Kinwood Holdings' John Deere feller buncher cuts and piles trees (Far Left) which are skidded to roadside with a grapple skidder (Top Right) and delimbed and topped with a stroke delimber (Bottom Right), after which they are loaded and trucked to the mill.

increasing capacity, chafed at the restrictions their quotas placed on them. As matters turned out, the Interior BC industry fared relatively well under the agreement, a circumstance that soon created huge rifts within the Canadian forest sector.

Throughout the period beginning with the 1991 dispute, much of Jake's time was spent negotiating with representatives of the US industry, including companies with which Lignum conducted business. This was not unusual, as there are many linkages between Canadian and US forest companies. At one point, for example, Jake's US counterpart in the negotiations, the chairman of the Coalition for Fair Lumber Imports, was Mack Singleton, president of New South Lumber in South Carolina, with which Lignum had a sales agreement. Both of them had to set aside their corporate interests to represent the industries of their respective countries.

From his competitors in BC and Canada, Jake learned a lot about the operation of other lumber companies functioning under the same rules and conditions as Lignum. It helped the company shape its own unique strategy. And his dealings with his American counterparts led to a fuller understanding of their position and how US and Canadian forest companies could work together to mutual advantage. For his work on behalf of the industry during this period, Jake was awarded the Order of British Columbia in 1997. In 2002 he was appointed a Member of the Order of Canada for his work with native people, his role in trade talks and for his philanthropic activities.

Throughout the 1990s, and in some respects aided by Jake's participation in the trade disputes, a new marketing strategy evolved. Known as "vendor-managed inventory," it evolved out of Lignum's long-standing reliance on the establishment and maintenance of close business and personal relationships with its partners, and from opportunities presented by the production of MSR lumber.

The new strategy involved bypassing the lumber wholesale sector and entering into partnerships with major US distribution companies. Lignum maintains an inventory of lumber, at guaranteed terms, in these companies' yards. In turn, they buy exclusively from Lignum. Some of them are small, private companies similar in outlook to Lignum, while others are major US forest companies.

The first of these arrangements was concluded in the early 1990s with Georgia Pacific. Not long after, a similar but larger deal was struck with a Boston-based company, Furman. It was later taken over by the much larger Boise Cascade, and the agreement continued. Boise Cascade is now Lignum's biggest customer. By the end of the decade, Lignum was involved in about 35 vendor-managed inventory arrangements.

Another type of marketing strategy was developed to take advantage of MSR opportunities. After identifying Phoenix, Arizona, as a major growth area in the US, Lignum sought out a truss manufacturer there that was focussed on its core business and had no interest in attempting to earn extra revenues playing in the commodity lumber market. Under an agreement with Schuck, a house framing company that manufactures its own trusses, Lignum maintains and manages at Schuck's plant in

Now we've got to the point where we've got 35 vendor-managed inventory locations. It's really what we are now. We are an inventory-er on an exclusive basis for a lot of distribution yards in the US. We take all the risk in the market, and they pay us a premium for that. And the premium more than makes up for the 30 days' risk that we're taking. We've made a lot of money doing that over the years. It's a pretty neat thing. A lot of other companies have figured out that this is the way to go. Now the whole industry talks about vendor-managed inventory and everybody's doing it. Canfor's doing it in a big way with Home Depot, in a much bigger way than Lignum could possibly do. We're talking maybe a billion feet. Georgia Pacific won't do business with guys now unless they're prepared to do some vendor-managed inventory carrying, so the mills, most Canadian mills now, have to do some of it, but nobody's in it like we are.

AARON ANDERSON

Site sensitive logging in the 1990s.
San Jose Logging's Timberjack 1216 forwarder (Top) loads precision-cut logs that have been felled,
delimbed, topped and bucked by a modern harvesting machine, such as the Valmet (Below)
owned by Jordef Enterprises, 1998.

Arizona a lumber inventory of the grades needed by Schuck. In addition, this facility supplies other truss manufacturers throughout the southwestern US. Lignum's first MSR sale was to Schuck, and 15 years later it is still the largest buyer through this sort of agreement. Similar partnerships were forged with truss makers in Denver, Las Vegas, Salt Lake City and Napa.

In the southern and southeastern US, where Canadian MSR is not in demand, Lignum began to develop reload centres for standard dimension lumber. Rail cars of lumber are shipped from Lignum's Williams Lake and other BC mills to these centres in Georgia, Texas, Missouri and Florida, where they are broken down into smaller, truck-load lots for sale in those regions.

Different types of relationships have been formed with other customers in the US and abroad. The basic strategy is to seek out, or respond to, potential customers whose business is compatible with Lignum's and dovetail an agreement to the mutual benefit of both companies. Each of them foregoes some short-term profits in their transaction for the sake of higher long-term profits produced by the synergies of their relationship.

This is not a simple way of doing business because it entails the forging of close, long-term relationships based on personal connections and trust. It means foregoing some of the fast money available in the lumber trade when demand jumps and supplies are short, investing that potential gain in a longer-term venture. When it was developed in the early and mid-1990s, that sort of arrangement ran against the traditions of the dominant North American wholesale lumber trade, which, to a great extent, was built on deal-to-deal relationships between commissioned salesmen.

The personal nature of this kind of market strategy has resulted in the formation of close relationships between Lignum people and others in the industry, especially in the US. This required and reinforced Jake's wider involvement in the industry on a provincial, national and continental scale. His increased participation in this end of the business was why a person like Conrad was needed to look after the ongoing management of the company. And Conrad's capabilities in that position underlay Jake's successes on the larger stage.

To some observers, Lignum appears to be overstocked with senior managers, considering the scale of its operations. On the other hand, its success during the 1990s demonstrated it is one of the most effectively managed. During this difficult period, most large companies were getting rid of senior managers at a rapid rate. Lignum, typical of its habit of marching to its own beat, acquired additional management capability during this period.

Another unique approach, involving a different type of relationship, evolved out of the decision by one of the Williams Lake mill employees, Gian Sandhu, to go into business for himself. In 1981, he had bought a small sawmill in Williams Lake, maintaining a minor business connection with Lignum. In 1987, he took advantage of a special form of timber sale established to encourage the manufacture of value-added forest products. He built a new plant, changed his corporate name to Jackpine Forest Products and entered into a long-term agreement with Lignum. Jackpine delivers its logs to Lignum, and Lignum supplies Jackpine with lumber which Jackpine remanufactures into higher value products—some of which Lignum markets.

Jackpine value-added products manufactured from lumber cut at Lignum's mill from logs supplied by Jackpine. Lignum markets some of Jackpine's production under one of the most successful value-added agreements in the province.

My initiation into the lumber industry started with Lignum, in '71 I think. I started with Jake's dad, as a lead hand. After 10 years I decided to go on my own and bought a small sawmill here with five employees. When I said, "Look, I'm going to go on my own," and when I gave my notice, Jake said, "No way, you're not going to go. That's not a threat to us, you can ask somebody else to run that tiny operation." So it took me six months to leave after I gave my notice. I had a small company I started in 1981, a small sawmill, just like every other sawmill, producing a truckload every second day. What brought the change in my heart? In 1978, if you recall, policy changes were taking place. Up to 25 percent of the cut would be taken from the primaries and put into the small-business program. I thought if I'm ever going to get a break in my lifetime, this is the one. If I've made a buck for Jake, then I should be able to make a buck for myself.

I built it up to 70 employees, with no timber. Made it into an 80 percent export company, cutting Japanese cuts. Baby squares that people on the Coast were doing out of hemlock I started doing out of Douglas fir. Doug fir as a species was still available in the Interior, a little bit more than is there right now. When log sources I developed started to disappear, I said, "Heck, I can't survive."

It depended upon log content, but the logs were not available. If the logs are not available, what do you do? The best thing is you go onto reman and get your raw material from the primaries so you're not a competitor to them. I was in the heart of the Interior, with five major corporations right here in town. The decision had to be to go into reman. That's the only decision I could do. Since I did that, I have never looked back.

❧

GIAN SANDHU

At this time, most primary producers in the provincial forest industry were reluctant to sign long-term supply agreements with smaller companies whose business was to remanufacture lower grades of dimensional lumber into higher value products. The bigger companies liked to play the markets and avoided long-term agreements for fixed volumes of lumber at an agreed-upon price schedule. The special timber sales were an attempt on the part of the government to give value-added producers some leverage by enabling them to trade logs for lumber. Even so, most primaries were unwilling to forge such agreements, hoping perhaps the remanufacturers would eventually disappear.

Lignum, in its usual contrary manner, embraced the idea. It had looked at the possibility of entering the remanufacturing business itself, tried a few experiments of that nature, and decided this was not part of its core business. Instead, it was determined that it would be better to let someone like Gian undertake that end of the business and stick with the basic undertaking of primary lumber production.

Lignum mill yard in the 1990s, with rough lumber waiting to be kiln dried.

Jackpine grew steadily under this regime during the 1990s, turning out a broad range of products such as finger-jointed lumber, joists, truss components, door and window components, furniture and specialty products. By the end of the century, Jackpine employed 225 people in two technically sophisticated Williams Lake plants, in the process establishing itself as Lignum's largest domestic customer. Lignum's sales division, in turn, sells some of Jackpine's products.

The close, complex nature of this relationship is unique in the BC forest sector, and the envy of most of the province's other value-added producers. It is a synergistic relationship that gives both companies higher levels of production than they could achieve separately, and higher returns on that production. The five-year agreements between them give both companies an added element of stability not enjoyed by most value-added or primary producers.

A similar relationship was struck with Westech Wood Products, which built a mill across the street from Lignum's Williams Lake operation. Westech buys planer mill trim ends, which it remanufactures into finger-jointed studs and fruit box components it sells in California. The Westech and Jackpine relationships, and other Canadian sales initiatives, were brought about in part by the 1996 Softwood Lumber Agreement, which limited Lignum's shipments, and future growth, in the US markets.

> *We don't try to be all things to all people. Some of the big companies are still pissed off over the fact there is a small business program where little guys are allowed to be remanners. Some of them still have a position that says, "Wipe them out, we know how to do reman."*
>
> *We tried reman, but that's not what we do. Again, it's a focus question. What we do is we make dimension lumber as efficiently as anyone in the world, but we're not very good when it comes down to fiddling with a two-by-three and making a piece of box stock from it. We don't have the mind set.*
>
> *This was Ailport. "What are we trying to do here? Let's let someone else make a buck."*
>
> *I think that has a lot to do with it. We really do believe it's okay for Jackpine to make a dollar or two, in fact we encourage that. If he's making a buck and we're working together, that's fine. It's the same as the deals with the First Nations, where we've been able to do some stuff.*
>
> *This may sound like an incredibly self-serving remark, but I think the heart of why all this works is that I'm not much of an accounting-oriented guy and I don't pay that much attention, after all these years, to exactly how much money we're making. I should pay attention right now when we might lose some. But if Gian Sandhu and ourselves can work out a relationship where he does something well, which he does, then good for him. We'll participate on both sides. We get logs on one end and we sell his stuff for him on the other.*
>
> *JAKE KERR*

Yet another form of relationship evolved in the 1990s out of the long-unresolved issue of aboriginal land claims. Throughout most of BC, treaties have never been signed with natives and increasingly, as native leaders took their disputes to court, it became apparent the legal basis of resource and land use in the province was uncertain. The forest industry, because of its extensive use of Crown forests, was particularly vulnerable to the uncertain investment climate this situation created.

Even before this time, beginning with Tim Kerr's tenure in Williams Lake, Lignum had begun development of business relationships with native communities in its operating areas. Band members from reserves throughout the region had worked for the company from its beginning, especially at Tatlayoko Lake, where employees were almost all native throughout the mill's 10-year operation. Under the direction of woodlands manager Brian LaPointe, after he was appointed to the position in 1988, Lignum began developing joint ventures with native bands in which independent forest companies were established with ownership shared equally between Lignum and the band councils.

The first of these ventures was Ecolink Forest Services, set up in 1990 in partnership with the Alkali Lake Band, with Johnny Watson as manager. Initially, Ecolink's primary business was silvicultural contracting—planting, spacing, seed collection and fighting insect attacks. By 1995,

Ecolink employed 60 people and had revenues of $2 million. Other bands in the region, watching Ecolink's success, objected to the Alkali Lake-based company operating in what they considered their territories. As a result, Ecolink's silvicultural work was scaled back and joint ventures commenced with other bands. In the late 1990s, Ecolink went into logging with a contract to harvest 60,000 cubic metres a year.

A second joint venture was established between Lignum and the Alexis Creek Indian Band. Chendi Enterprises engages in contract road construction, silvicultural work and logging. It has been a successful and, relatively, stress-free partnership. In addition, a number of smaller joint ventures have been entered into with various bands in the Williams Lake area.

People say, "How can you have these partnerships with First Nations? Everyone knows they don't work, they're sort of a charitable thing."
Bullshit. These are very profitable companies and they're run on a business basis.

JAKE KERR
in Business in Vancouver, 1998.

To some extent, these joint ventures between Indian bands and private companies are hampered by the apparent inability of governments and native leaders to resolve land claims. Settlement of these claims, it is hoped, will clarify the relationships between companies such as Lignum and Indian bands in whose traditional territories they operate. In the meantime, uncertainty over forest resource rights makes relationships between the partners in these joint ventures, even successful ones such as those in which Lignum is involved, more difficult than other types of business relationships.

Beginning shortly before the 1991 provincial election, public concern developed about the management of forests in the Cariboo-Chilcotin. Some of this concern was generated by forest companies such as Lignum increasing their logging operations west of the Fraser River to salvage beetle-killed pine. Under policies established in the wake of the 1945 Sloan Royal Commission, the province's forests had been managed to sustain timber supplies. Other policies were intended to protect non-timber forest values in areas harvested.

Second-growth forest spaced and pruned by Lignum in 1994. By the late 1990s, Lignum was spacing and pruning almost 1,500 hectares of young forest a year.

By the early 1990s, two things had become apparent: the timber supply could not be sustained under these policies, and non-timber values, such as wildlife, water, scenic and other attributes, were not being adequately cared for. A politically effective preservationist lobby developed to press for the protection of large areas of forest land, and the exclusion of these lands from industrial use. On the Coast, enormous confrontations developed between environmental protectionists and forest companies.

The Harcourt government established a process, the Commission on Resources and the Environment (CORE), to try to reconcile these conflicts. One of CORE's specific tasks was to devise a land-use plan for the Cariboo-Chilcotin upon which everyone in the region could agree. Lignum participated fully in this process, which involved 62 full days of meetings spread over many months.

The various interests involved formed two coalitions. The environmental groups formed one, and the economic interests, such as the forest industry, formed a second group known as the Cariboo-Chilcotin Community Coalition. Lignum was a member of the latter. One of the economic coalition's objectives was to maintain an annual timber harvest of 7 million cubic metres in the region.

When the two sides failed to agree on a plan, CORE produced one. No one on either side of the debate liked it, and everyone was disappointed their efforts had led to naught.

One of CORE's staff consultants through much of this process was Bill Bourgeois, a former forester with MacMillan Bloedel. In this capacity he was able to observe how forest companies in various parts of the province responded to new ideas about forest management raised during negotiations. Very few of them, he noted, were seriously interested in undertaking the kind of forest management he thought was required to meet industry's need for increased timber supplies and the concerns of other forest users for sustaining non-timber forest values.

One of the few companies to express an interest in engaging in increased levels of forest management was Lignum. After a series of discussions with Jake on the subject, Bill was appointed vice-president of forest policy, with a major part of his job being to help represent Lignum in the growing forest policy debate.

He and Jake drew up a paper, which Lignum published, in the summer of 1994, as a set of ten recommendations for improving the CORE plan. They offered a means of maintaining, and even increasing, economic use of the forests, while meeting the needs and wishes of others.

A key factor in what followed was Lignum's faith in the importance of relationships and of communicating with whomever the company does business. At this time the forest industry and environmental interests were squared off into two opposing camps. Jake encouraged Bill to engage in informal discussions with some of the leading environmentalists.

Soon, the two interest groups resumed meeting outside the CORE process. When they began to make some headway, the government provided a mediator who assisted in drafting a "made-in-Cariboo" plan. It was given to the government, which quickly adopted it as the official land-use plan for the region.

The near-failure of the Cariboo-Chilcotin CORE process was symptomatic of the Harcourt government's well-meaning but awkward attempts to deal with festering forest policy issues ignored by previous governments. The result was increased conflict and confusion in the forest sector. In 1996, when Harcourt was hounded from office and replaced by the NDP's more ideologically doctrinaire Glen Clark, confusion was compounded and investor confidence in the industry further undermined. Through the 1990s, Jake was one of the leading spokesmen for industry with government, not always a pleasant or productive task.

In its early attempts to deal with the conflict its policies were generating within the forest sector, the Harcourt government had set up a forum of senior bureaucrats and representatives from industry to develop new policies. Jake had learned from Leslie that no matter how much one disliked a particular government, it was the government the people had elected, and had to be dealt with. He was appointed to the forum, known as the Forest Sector Strategy Committee. It

Fertilizing a young forest with urea. By the late 1990s, Lignum was aerial fertilizing about 1,750 hectares a year

was instrumental in establishing a new forest-sector funding agency, Forest Renewal BC (FRBC), in 1994. Jake was appointed to the FRBC board of directors, where much effort was required to fend off attempts by the Clark government to channel funds to non-forest expenditures.

The height of the Clark government's forest-sector folly was probably its instructions to the industry in 1997 to create 22,000 new jobs within the next five years, or risk losing cutting rights to Crown timber. Soon after, the fortunes of the forest industry hit another decline and the jobs scheme was quietly forgotten in the following months as employment in the industry declined. It was difficult for many companies to even survive this period, let alone prosper and grow.

In spite of these difficulties, Lignum continued to evolve, responding to opportunities as they arose. One of these opportunities appeared in 1995, when the government asked for proposals to develop new, experimental forms of tenure to practise sustainable forest management.

In BC, as in most other Canadian provinces—but unlike the rest of the developed world— the provincial government owns the forests. Through various forms of leases and licences, timber from these forests has been made available to the private sector for the manufacture of lumber, pulp and paper, and other timber products. For the most part, forests harvested or burned have been restocked by government or industry, so the province's forests are intact.

What has not developed under this tenure arrangement is the investment of private capital into the forests to improve their health, to increase the quantity and quality of timber grown, and to care for non-timber values, such as wildlife and water. Elsewhere, where private forest property rights are more extensive and more secure, much higher levels of forest management are practised and, among other benefits gained, timber supplies have been significantly improved and other values protected. Thus far, these opportunities have not existed in most of Canada, including BC.

An opportunity to manage the forests in Lignum's operating area more intensively arose in 1995.

The government's proposal came in the wake of a document published by a national forest coalition of governments, industry, environmentalists and other interested parties, known as the Canada Forest Accord. Its stated goal is "to maintain and enhance the long-term health of our forest ecosystems, for the benefit of all living things both nationally and globally, while providing environmental, economic, social and cultural opportunities for the benefit of present and future generations."

Using this statement as a guide, Lignum submitted a proposal which, in 1997, was accepted as the first of six Innovative Forestry Practices Agreements (IFPA) located throughout the Interior. Lignum's agreement covers 450,000 hectares of forest land and is similar in many respects to the management concept proposed by Leslie before the Sloan commission 40 years earlier.

In other respects the IFPA is a very nebulous concept. It has highly laudable goals, which include: development of socially acceptable sustainable forest management plans and practices; increasing the timber supply; implementation of a performance-based approach to forest management; communication of results so as to influence forest management around the province; conservation of environmental values; the promotion of tenure reform; and improved knowledge of forests.

The potential gains from this approach are enormous. Acquiring the ability to sustainably manage all forest values would give any company operating under such a system strong support from the public and preclude the type of conflicts which have marked the past two decades. It also offers the possibility to increase the timber yields substantially. This opportunity is consistent with Lignum's overall management philosophy of improving its business prospects by making the best of what it has, rather than growing through the acquisition of other companies.

On the other hand, the agreements concluded were for ten-year periods only. There are no assurances of renewals or bankable guarantees that companies such as Lignum will receive the benefits of any success they have in increasing timber yields. The plan carries a lot of risks and, at this stage, no guarantees. If the IFPAs are not permitted to evolve and provide incentives for significant investment in the forests they cover, they will fail.

In 2000, Lignum's management proposals for the IFPA were accepted. The company's new woodlands vice-president, Dave Peterson, began restructuring its forestry department, which was provided with a new space-age addition to the Williams Lake office building. Initial work has focussed on gathering inventory data and planning intensive management programs. Perhaps the major change is that Lignum has begun to redefine the scope of its woods operations, from the harvesting of timber to the management of forests.

Consistent with this development, and to provide a third-party evaluation of its forest

Williams Lake near the end of the 20th century, with the old Lignum mill site in the centre of the photo, and the present mill site upper right.

We're almost to the point of being able to project future timber yields from increased forest management. It will probably take another year or two at the most. We now have done the work on what we call "vegetation inventory," which is equivalent to a timber inventory, and a bunch of other vegetation. So we've done that, we have 175 permanent sample plots established. This is the first year where we're starting to re-measure some of those that we put in over the years. So we start the five-year re-measurement on some.

That will provide us with some information.

We've done site index work to know what the productivity is, then a development of yield curves. In about a year we'll have all that completed. We know where all our roads and landings are so we know how much of the land base has been taken out of production. We know the terrestrial ecosystem, the ecological mapping, we know that. All our streams for fish and fish habitat have been inventoried. We have all the topography, albeit that a lot of the area is flat anyway.

The recreational inventory, at a reconnaissance level, the government already had, so there's no need to do anything more there. We've done all the mapping on the visually sensitive areas along travel corridors. We know what the plants are around those areas so we know how much timber we can take up and maintain the visual quality. So we're in pretty good shape with that.

We haven't done the number crunching. We haven't done the base-case analysis, those kinds of things but a couple of years from now, we'll be okay. One area that we're still deficient in is what we call "managed stand plots," so we can know the growth projection of a natural stand and what happens under various treatments, what's the projection there. We haven't got that data, we'll just have to use government data at the present time. We are hoping we will find some funds to put in those plots. Then the base stuff will be in place.

That's the only area we're still deficient in.

We're doing a lot of other research, different things like seral-stage distribution, what are the structures in an uneven-aged stand that make it considered old growth, those kinds of things, some interesting stuff. Fire-return intervals, and that sort of thing. So we're pretty pleased with what the guys have been doing.

BILL BOURGEOIS

practices, in 1998 Lignum applied for certification of its operations from three of the many plans available worldwide, two of which were achieved by the end of the century.

Over the past two decades, in spite of bug attacks, unstable governments, environmental protests and other difficulties, Lignum has maintained its policy of constant improvements to its mill, reinvesting money steadily to maintain its efficiency in the face of shifting market conditions and fluctuations in the timber supply.

Since 1980, well over $40 million has been invested in the mill complex. While this included some major additions—a second Chip-N-Saw and the MSR system in 1986, an edger optimizer in 1987, and a chip plant in 1996, for example—much of this reinvestment was for upgrades to the existing setup.

This constant fine tuning of the mill, under the direction of Rob Fraser and Lignum's technical superintendent, Jim Thomson, is a major achievement. It is not as dramatic as a major rebuild or addition, but it is far more difficult to pull off without disrupting production. In 2000, in spite of utilizing the lowest quality log supply in the province, the mill was ranked by Price-WaterhouseCoopers as first in cost competitiveness among Cariboo mills.

When a second major mountain pine beetle attack developed in the late 1990s, Lignum was able to respond quickly. In 1999, about 95 percent of the company's harvest on Crown land was beetle affected wood, either from the earlier infestation or the new one. By that time, 4.5 million cubic metres of pine in Lignum's operating area—a 15-year supply for the Williams Lake mill—was affected. The entire logging program was focussed on the removal of infested or susceptible trees, and the mill was geared to processing this timber.

Lignum's Williams Lake mill in the 21st century.

Early in 2002, with a new government in power in BC and lumber markets shrinking, the

Canadian lumber industry found itself yet again under attack by the US forest industry. A countervail tariff, combined with additional duties imposed for alleged dumping of timber in the US, levied duties of 32 percent on Canadian lumber sold in the US. Most of Lignum's US sales were subject to these duties. Jake, as co-chair of the BC Lumber Trade Council, was once again in the thick of negotiations.

Still, with two-thirds of BC's coastal lumber industry shut down, and most Interior companies implementing temporary closures to reduce lumber inventories, he visited the Williams Lake mill to assure employees that Lignum's customers continued to buy the mill's products and that no closures were contemplated. Lignum was one of only a handful of companies in the country able to make that decision.

By the beginning of the new century, Lignum's Williams Lake mill complex was one of the most efficient producers of premium forest products in the business. Its operations have never closed because of slow markets.

The Lignum sales division has transformed itself, and led the industry, from a wholesaler of dimensional lumber into a marketing company that constantly works more closely with the end-users of its products. In the late 1990s, for example, it entered into a supply agreement with Centex, the second-largest home builder in the US.

With its marketing operations able to maintain steady sales of everything its mill produces, and more, and its mill capable of converting the lowest value timber in the province into high-value lumber, Lignum is also poised—when the long-awaited reform of provincial forest policies takes place—to lead the industry into its next phase, the growing of an enhanced timber supply.

> *We do a lot of things that are not in Lignum's best interests in the very short term. We are always driven by things that are in Lignum's and the customer's best interests in the long term.*
>
> CRAIG STUART

At the beginning of the 21st century, Lignum had realized, or was in the process of realizing, the vision articulated by Leslie Kerr when he appeared before Chief Justice Sloan's Royal Commission in 1956. His predictions about the evolution of the industry turned out to be quite accurate. His ideas about how the company should proceed, especially considering the constraints imposed by provincial forest policies, have been largely fulfilled. If he had lived to see it, he would recognize and be pleased with what his successors at Lignum—his sons and the children of many of his contemporaries—built on the foundations he laid.

ACKNOWLEDGEMENTS

This book is a product of many people's efforts and goodwill. Most of those interviewed are named in the excerpts taken from these interviews. Others, whose time talking with me is much appreciated, are Rob Fraser, Brian La Pointe, Harvey Arcand, Roger and Darren Getz, Andy Chelsea and Johnny Watson.

Bill Phillips at the *Williams Lake Tribune* was helpful with information and photos, as were the staff at the Williams Lake Library. Mary McRoberts, author of an excellent master's thesis and articles on Lignum, was most generous in her assistance.

Roy Astley, the company's unofficial historian, deserves recognition for ensuring the early history was not lost. Jim Thomson, Lignum's technical superintendent, patiently supplied detailed technical information. Al Dupilka, of Quesnel, is to be thanked for his generosity in sharing his knowledge of the early history of the industry in that city, and of the Cariboo forest industry in general. And Peter Stanley is to be praised for conveying questions and answers.

I especially would like to thank Sidney Eger, Wendi Struthers and Betty Engemoen for their enthusiastic assistance and patience in helping with the preparation and production of this book.

Ken Drushka
Vancouver, 2002